An Introduction to Ecological Modelling

Putting Practice into Theory

METHODS IN ECOLOGY

Series Editors
J.H. LAWTON FRS
Imperial College at Silwood Park,
Ascot, UK

G.E. LIKENS
The New York Botanical Garden,
New York, USA

METHODS IN ECOLOGY

An Introduction to Ecological Modelling

Putting Practice into Theory

MICHAEL GILLMAN

Department of Biology, The Open University
Walton Hall, Milton Keynes

ROSEMARY HAILS

NERC Institute of Virology and Environmental Microbiology
Mansfield Road, Oxford

Blackwell
Science

© 1997 by
Blackwell Science Ltd
Editorial Offices:
Osney Mead, Oxford OX2 0EL
25 John Street, London WC1N 2BL
23 Ainslie Place, Edinburgh EH3 6AJ
350 Main Street, Malden
    MA 02148-5018, USA
54 University Street, Carlton
    Victoria 3053, Australia

Other Editorial Offices:
Arnette Blackwell SA
    224, Boulevard Saint Germain
    75007 Paris, France

Blackwell Wissenschafts-Verlag GmbH
    Kurfürstendamm 57
    10707 Berlin, Germany

    Zehetnergasse 6
    A-1140 Wien
    Austria

First published 1997

Set by Setrite Typesetters, Hong Kong
Printed and bound in Great Britain
at the Alden Press Ltd,
Oxford and Northampton

The Blackwell Science logo is a
trade mark of Blackwell Science Ltd,
registered at the United Kingdom
Trade Marks Registry.

DISTRIBUTORS

Marston Book Services Ltd
PO Box 269
Abingdon
Oxon OX14 4YN
(*Orders*: Tel: 01235 465500
        Fax: 01235 465555)

USA
Blackwell Science, Inc.
Commerce Place
350 Main Street
Malden, MA 02148-5018
(Orders: Tel: 800 759 6102
        617 388 8250
        Fax: 617 388 8255)

Canada
Copp Clark Professional
200 Adelaide Street, West
3rd Floor, Toronto
Ontario M5H 1W7
(*Orders*: Tel: 416 597 1616
        800 815 9417
        Fax: 416 597 1617)

Australia
Blackwell Science Pty Ltd
54 University Street
Carlton, Victoria 3053
(*Orders*: Tel: 3 9347 0300
        Fax: 3 9347 5001)

A catalogue record for this title
is available from the British Library

ISBN 0-632-03634-6

Library of Congress
Cataloging-in-publication Data

An introduction to ecological modelling:
    putting practice into theory/
    [editors], Michael Gillman, Rosemary Hails.
            p.   cm.–(Methods in ecology)
    Includes bibliographical references and index.
    ISBN 0-632-03634-6
    1. Ecology–Mathematical models.
    2. Population biology–Mathematical models.
    I. Gillman, Michael. II. Hails, Rosemary.
    III. Series.
    QH541.15.M3I53   1997
    574.5'01'51–dc20   96-35515
                                            CIP

Contents

The Methods in Ecology Series

The explosion of new technologies has created the need for a set of concise and authoritative books to guide researchers through the wide range of methods and approaches that are available to ecologists. The aim of this series is to help graduate students and established scientists choose and employ a methodology suited to a particular problem. Each volume is not simply a recipe book, but takes a critical look at different approaches to the solution of a problem, whether in the laboratory or in the field, and whether involving the collection or the analysis of data.

Rather than reiterate established methods, authors have been encouraged to feature new technologies, often borrowed from other disciplines, that ecologists can apply to their work. Innovative techniques, properly used, can offer particularly exciting opportunities for the advancement of ecology.

Each book guides the reader through the range of methods available, letting ecologists know what they could, and could not, hope to learn by using particular methods or approaches. The underlying principles are discussed, as well as the assumptions made in using the methodology, and the potential pitfalls that could occur—the type of information usually passed on by word of mouth or learned by experience. The books also provide a source of reference to further detailed information in the literature. There can be no substitute for working in the laboratory of a real expert on a subject, but we envisage this Methods in Ecology Series as being the 'next best thing'. We hope that, by consulting these books, ecologists will learn what technologies and techniques are available, what their main advantages and disadvantages are, when and where not to use a particular method, and how to interpret the results.

Much is now expected of the science of ecology, as humankind struggles with a growing environmental crisis. Good methodology alone never solved any problem, but bad or inappropriate methodology can only make matters worse. Ecologists now have a powerful and rapidly growing set of methods and tools with which to confront fundamental problems of a theoretical and applied nature. We hope that this series will be a major contribution towards making these techniques known to a much wider audience.

John H. Lawton
Gene E. Likens

Preface

To do ecological modelling we need to use some mathematics. The problem with mathematics is that many people have had a bad experience of it at an early age. So many students say 'I can't do maths' or 'I'm a biologist'; as ecologists we can't escape... The difficulty with teaching ecological modelling is that as soon as the ratio of equations to text reaches a critical point the shutters come down and no further progress is made. In this book we have tried to avoid that by immersing the equations in the text and explaining all the key mathematical steps. Perhaps an advantage we have is that neither of us has trained as a mathematician—we've started as ecologists and continue to practice as ecologists. However, we have found mathematical modelling useful and stimulating and hope to convey that to as wide an audience as possible.

The aims of this book are: (1) to provide an overview of the structure and applications of population and community ecological models; (2) to demonstrate how to construct, use and test ecological models using examples with minimal mathematical content, and (3) to encourage and facilitate readers to develop their own models.

To understand this book you will need no more than basic high school maths; particularly the ability to manipulate equations—GCSE (UK) or Junior High (US) level. There are several areas of maths where some prior knowledge (even in the distant past) would be helpful; e.g., differentiation, matrices, logs, but we do provide boxes on these and other mathematical topics. The mathematics in the boxes is provided on a 'need-to-know' basis, to cover the bare minimum of knowledge, focusing on the principles with a few examples. The boxes are not intended as a substitute for a standard mathematics course or text but rather as a set of notes to help readers in working through this text and to direct them to relevant material in supporting mathematics texts.

It is intended that the examples and methods used in this text will provide a bridge between application and theory and allow those readers with little or no present understanding of modelling to access the ecological modelling literature. The text is designed to provide a progression of mathematical techniques, with questions at the end of some of the sections and bold terms to flag key ecological modelling concepts or results.

There are a large number of texts which have addressed ecological modelling—these have usually been written at a high mathematical level, often by authors with a strong background in mathematics. Whilst these texts are rich in ideas and have contributed to the development of modern theoretical ecology, they necessarily begin at a relatively high mathematical level and therefore the diversity and subtlety of their arguments may be lost on many ecologists. The one exception is Wilson and Bossert's *Primer of Population Biology* published in 1971 which took an interactive approach coupled with the low mathematical threshold adopted in this text.

For a further exploration of the structure and output of some ecological models visit the Blackwell Science home page http: //www.blackwell-science.com/products/ ecolmods and look at the catalogue entry for this book which will have a link to a number of worked models.

M. Gillman
R. Hails

Acknowledgements

We are grateful to John Latto, Trevor Williams and Byron Wood for reading parts of the book. Christine Randall (Open University) helped with equation editing and provided general secretarial support. We thank Jonathan Silvertown for providing access to unpublished data on *Cirsium vulgare* and Phil McGowan and Ding Chang-qing for unpublished data on Cabot's tragopan. The Open University Ecology Course team (Jonathan Silvertown, Irene Ridge and Dick Morris) contributed to a lively debate on the level of mathematical knowledge appropriate to undergraduates studying ecology. Two anonymous referees made helpful comments on much of the book. Finally, we are extremely grateful for the patience and assistance of the Blackwell Science staff, especially Susan Sternberg, and the encouragement of the series editors.

CHAPTER 1

Introduction: themes of ecological modelling

1.1 What is an ecological model?

Ecology is concerned with the interactions between an organism and its environment (both abiotic and biotic, i.e. other organisms) and the consequences of these interactions, including the change in numbers of individuals in populations (single species) or communities (multi-species). In this book we focus primarily on population and community models (see Section 1.2). For example, in population ecology a major aim is to describe the change in the numbers of individuals in a population over time and/or space (collectively referred to as **population dynamics**, Fig. 1.1) arising out of interactions such as death by predation.

An ecological model must be able to describe this change in numbers to varying degrees of accuracy and generality (see Section 1.3). Such models are phrased in mathematical language. There are three reasons for this: (1) brevity and formality of description; (2) manipulation of the model and (3) the discovery of emergent properties not apparent from non-mathematical reasoning. Thus, instead of the clumsy statement that the number of individuals in a population next year (N_{t+1}) is given by the number this year (N_t) minus the number of deaths (d) and emigrants (e) from the population plus the number of births (b) and immigrants (i) we can write:

$$N_{t+1} = N_t - (d + e) + (b + i) \tag{1.1}$$

This equation can then be manipulated as can any mathematical equation. This leads to a flexibility which goes well beyond a conceptual model phrased in ordinary language. Furthermore there are emergent properties and results of mathematical models, typified by the appreciation of chaos in population dynamics (see Section 1.4 and Chapter 2), which are difficult or impossible to predict from non-mathematical reasoning. It is no coincidence that many of the clearest thinkers and greatest contributors to the development of ecological theory have been those with a thorough grounding in mathematics. This is largely owing to the fact that mathematics requires a precise and logical formulation and therefore forces the user to explicitly state particular rules, e.g. defining density dependence (Chapter 2) in mathematical terms is much easier and more precise than in ordinary language. For any mathematician reading this paragraph these statements are simply the *raison d'être* of mathematics. For sceptical ecologists and newcomers to mathematical modelling we hope that they will be convinced by the studies presented in this text.

(a)

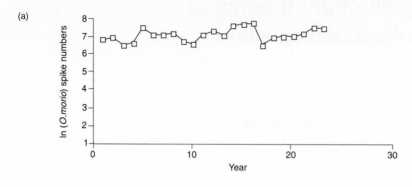

(b)

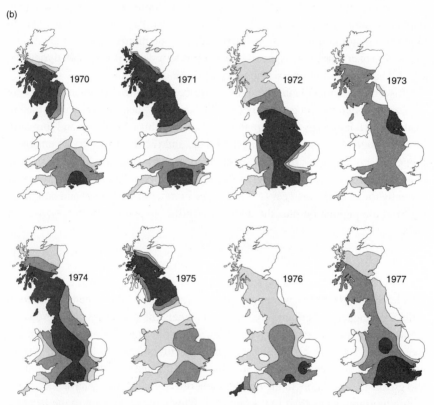

Fig. 1.1 Examples of (a) temporal and (b) temporal and spatial population dynamics. (a) Shows the change over time in a population of the green-winged orchid *Orchis morio* at one locality (Gillman & Dodd, in press). In (b) the change in spatial distribution in Britain of elder aphid (*Aphis sambuci*) is depicted between 1970 and 1977 (Taylor & Taylor 1979). The density of shading indicates the local population density.

1.2 Applications of ecological models

There are few subjects in ecology which cannot benefit from an appreciation of mathematical models or cannot be couched in mathematical terms. Such an appreciation may occur both at the analytical or conceptual level or arise out of simulations (defined in Chapter 2). Ecological science can be divided into four related subject areas, each with its own range of applications (Table 1.1). For example, mathematical models may be used to explore the spread of a species in response to global climate change and describe the rate at which this occurs by linking local population dynamics and migration. The application of models may have profound economic implications, such as models which indicate the likelihood of success of a biological control agent (perhaps dramatically reducing the economic and environmental cost of pesticide application) or suggest the efficacy of a programme of drug treatment.

Although in this book we concentrate on population and community ecology it is necessary to refer occasionally to physiology and ecosystems. This is because the classification in Table 1.1 hides a great deal of interdependence: thus, the study of population structure and dynamics is necessarily concerned with elements of community ecology (within and between trophic level interactions). Likewise, questions in community ecology such as 'is predation a major structuring force in communities?', requires an understanding of population processes. It is perhaps in population ecology that mathematical modelling has had its greatest impact and it is therefore a major theme of this book. Indeed, it is difficult to venture too far into population ecology without an understanding of mathematical modelling. The application to population dynamics also has a long and fascinating history going back to the Fibonacci series (found by starting with the two numbers 1, 1 and adding the previous two terms to get the next term, thus 1, 1, 2, 3, 5, 8, 13, ...) which was apparently a model (albeit somewhat naive) of rabbit population dynamics developed in the thirteenth century (Hoppensteadt 1982). Much of the development of population models was due to a desire to understand human

Table 1.1 The branches of ecology and examples of applications of mathematical modelling.

Subject areas	Range of applications
Physiological ecology	Foraging, digestion rates, allometric growth, rates of translocation and transpiration
Population ecology	Biological control, harvesting of species, spread of invasive species, including disease amongst humans
Community ecology	Community stability and diversity, coexistence and species richness
Ecosystem ecology	Nutrient cycling, effects of global climate change

population dynamics (Malthus 1798, Verhulst 1838, Pearl & Reed 1920, Lotka 1925, Volterra 1926, 1928), elements of which are discussed in Chapter 3.

1.3 Simple and complex models

The complexity of ecology is both its fascination and frustration—it may involve many individuals of various species interacting with a variety of abiotic and biotic factors which themselves may be affected by sets of other factors. All of these factors are likely to change in space and time, often unpredictably. How then can we begin to model these systems? There are two extreme approaches which have been described by various authors, e.g. Maynard Smith's (1974) distinction between practical 'simulations' for particular cases and general 'models', May's (1973a) distinction (following Holling 1966) between detailed 'tactical' models and general 'strategic' models and Levins' (1966) 'contradictory desiderata of generality, realism and precision'. These approaches have received some criticism but, at the very least, they provide a starting point in understanding our motivations for modelling.

At the 'tactical' end of the spectrum we attempt to measure all the relevant factors and determine how they interact with the target population or community. For example, in producing a model of change in plant numbers with time we might find that the plants are affected by 12 factors, including summer and winter rainfall, spring temperature and herbivores. We obtain this information through field observations and experiments. All the information is combined into a computer program, initial conditions are set (e.g. the number of plants at time 1), values for the variables entered (e.g. level of summer rainfall) and the model run. With increasing computer processing speed and memory it has been possible to incorporate more and more detail with models exemplified by the new generation of individual-based models (see Judson 1994 for a review of these) and the huge ecosystem simulations which were originally developed in the late 1960s as mainframe computers became widely available. After a certain number of generations or time periods the number of plants is recorded from the program output.

Now comes the tricky part. We have produced a 'realistic' model in the sense that it mimics closely what we believe is happening in the field. However, we do not really know why it produces the answer it does. The model is intractable (and perhaps unpredictable) owing to its complexity. Tweaking a variable such as rainfall may radically change the output but we do not know why. In other words we have created a black box which receives ecological variables and spews out population dynamics. The main value of such a model is that it can speed up natural processes so that we do not have to wait 100 years to see how the plant population will (possibly) change, assuming extrinsic factors remain the same or change in a predictable manner. But, we are often no nearer to the mechanism(s) driving the dynamics. At this stage we have two options (other than scrapping the model). The first is to alter the values of the variables systematically and see how the output responds. This is perhaps best undertaken after the second option which is to strip down the model to its (most) statistically significant components.

In statistical terms we have a dependent variable, e.g. plant change over time, and a series of explanatory variables, e.g. temperature and rainfall. These variables explain a certain percentage of the variance in the dependent variable. The more explanatory variables which are included in the model, the greater the total amount of explained variance in the dependent variable. But the addition of an extra variable may only add a small (and possibly statistically non-significant) increase in explained variance. We therefore need to remove all the explanatory variables which do not provide any significant increase in percentage variance explained. We may then be left with, for example, three variables which explain a total of perhaps 70% of the variance. (See Section 1.4. and Chapter 2 for a discussion of the unexplained variance.) This is likely to be more tractable than our original model with 12 explanatory variables. We can now manipulate these variables (perhaps coupled with experiments in the field or microcosm) and explore their effects. We will refer to this as the simplest realistic model. Not all ecological models can be simplified in this way but it is important to try to formulate models which can be simplified according to objective criteria, such as statistically significant gain or loss of explanatory power.

From the 'strategic' end of the spectrum we can create a model which is so simple that it is known to be unrealistic. What is the value in this? Here the objective is rather different to the simplest realistic model. We are using mathematical modelling as a way of formalizing generalizations about the ecological system(s) of interest. The model is not derived out of consideration of one particular example (as above). The Lotka and Volterra predator–prey models in Chapter 3 and the Levins metapopulation model in Chapter 6 fall into this category. Such models are 'strategic' as defined above. In many ways they are the most important types of model as they lie at the heart of the 'realistic' models described above (as will be seen throughout the book). If we do not understand the mode of operation of strategic models then we can never understand why the particular realistic models do what they do. Thus, the phenomenon of population cycles (e.g. in Chapters 2–4 and 7) cannot be easily understood without reference to strategic models. These models also have the advantage of being relatively simple and often based on one or two equations so that they can be easily manipulated. These equations also appear in a variety of guises and it is fun to try to spot the strategic equation (and its provenance) in an ecological modelling study.

1.4 Discrete and continuous time, stochastic and deterministic processes

In mathematical models of population or community change over time there are two ways of representing time which have important implications for the way the populations or communities are modelled. In the first case time may be considered as **continuous**, so that, in theory, it can be divided up into smaller and smaller units. In the second case, time is considered to be **discrete** and indivisible in units of, e.g. years. The first case is appropriate to populations of individuals with asynchronous and continuous reproduction (birth-flow models, Caughley 1977) whilst the second case is appropriate to populations composed of individuals with synchronized reproduction at regular time intervals (birth-pulse models, Caughley 1977).

Mathematical equations describing change in continuous time are **differential equations** (Chapter 3) whilst equations describing change in discrete time are **difference equations** (Chapter 2). The extremes described by these two types of equation may not be encountered in the field. Most environments have some form of seasonality, even in tropical habitats. Within the breeding season, there may be one reproduction event per (female) individual, in which case a difference equation is the most appropriate description, describing change between breeding seasons. However, there may be several, possibly overlapping, reproduction events in a breeding season, in which case a mixture of difference and differential equations may be suitable.

In a **deterministic** world everything should be predictable. If population dynamics are deterministic we should be able to predict the population size at time t and/or position i given a knowledge of the processes (described by mathematical equations) underlying the dynamics. This notion of the deterministic world has been upset by the realization that apparently random dynamics, known as chaos, can be produced under strictly deterministic conditions (Fig. 1.2; chaos in ecology is reviewed by May (1974, 1976) and discussed in Chapter 2 and Box 2.5).

No ecological system is purely deterministic. There are always some unexpected or unpredictable events, such as storms, which may have a strong effect on the dynamics of the target species. These may be entirely random in their occurrence in which case we refer to them as **stochastic** events. However, the randomness of these events may depend on the time-scale used. Thus, while it is difficult to predict the probability of storms from day to day, we may be much more certain about the occurrence and even quantity of such events from month to month. So an unpredictable (and effectively random) event at one time-scale may be predictable (and therefore deterministic) at another time-scale. The same effect can be expected of analyses at different spatial scales. There will also be inaccuracies in samples of plant or animal densities. If we take the simplest realistic model description a little further we can describe the extrinsic random events and the sampling error plus the non-significant deterministic factors as the unexplained variance (or simply noise or error) in the system. Chapter 2 develops this theme with respect to probability of extinction.

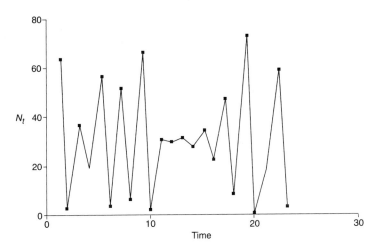

Fig. 1.2 Chaotic dynamics produced by a simple deterministic equation $[N_{t+1} = \lambda N_t(1 - N_t/K)]$ in which N_t is the population density at time t. See Chapter 2 for further details.

The apparent importance and ubiquity of random events has led some ecologists and mathematical modellers to dismiss the deterministic approach, or at least downgrade its importance below that of stochasticity. Ideally, one would attempt to combine both approaches in any modelling application (as emphasized by Renshaw 1991 and illustrated in Chapter 2 of this text). Overall we have tended to emphasize the deterministic aspects of modelling for two reasons. First, the analysis of stochastic processes is generally more complex than deterministic ones and therefore less suited to an introductory text. Second, we prefer to think of a primarily deterministic world clouded by stochasticity where the use of the simplest realistic model is appropriate. Stochastic population processes are only likely to override deterministic ones under particular circumstances, e.g. when population sizes are very small or when catastrophic unpredictable events occur (although ecological catastrophes such as earthquakes, volcanoes and hurricanes are usually spatially predictable if not temporally predictable).

1.5 Testing ecological models

The testing of ecological models is, in principle, straightforward. The output of a model of, say, plant population dynamics should be compared to the observed dynamics in the field. The better the description of the observed dynamics, measured by some objective statistical criterion, the better the model. In practice there are several problems. First, the values for the model components, such as birth rate, may have been taken from the field population against which the predicted dynamics are to be compared and so it is inevitable that model and field will show some agreement. Ideally, estimation of model components should be undertaken independently from model testing. Second, there are rarely sufficient field data for statistical testing, particularly where the dynamics are possibly cyclical or chaotic. In these cases we may need a run of population data covering several lifetimes of an average investigator. Often the best we can hope for is that the components of the model, such as birth rate or migration rate, are ecologically realistic and based on careful field measurements. Additionally, dependent on the aim of the model, field or microcosm experiments may complement the modelling. There is a two-way trade between field experiments and ecological models. Not only can experiments be used to parameterize and test the predictions of models and suggest the construction of new models, but also models can be used to indicate the design of field experiments. We consider a field experiment involving the population dynamics of spear thistle (*Cirsium vulgare*) in Chapter 4.

1.6 Stability and equilibrium

The **equilibrium** or steady state is defined as the state, e.g. population density, to which or around which a population will move. **Stability** is related to equilibrium in that it describes the tendency of a population or community to stay at or move towards or around the equilibrium. Both phenomena can be conceptualized by a ball in a cup (Fig. 1.3), i.e. a physical model. They are both investigated by **perturbation** or displacement of the ball and observation of its movement after perturbation. Initially the ball is at rest at an apparent equilibrium. Movement of the ball from position 1 to position 2 (Fig. 1.3a) and release of the ball (Fig. 1.3b) reveals position 1 to be an equilibrium position (we would not have known this by leaving the ball at position 1) which is locally stable from position 1 to position 2

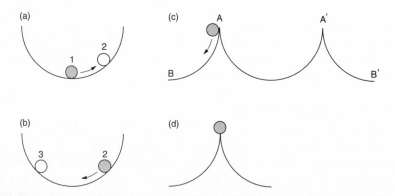

Fig. 1.3 Stability and equilibrium illustrated by a ball in a cup. (a) Displacement of ball from (apparent) equilibrium at position 1 to position 2. (b) Release of ball from position 2 (or equivalent position 3). (c) Displacement of ball beyond local stability boundary at A or A'. (d) Unstable equilibrium.

(and from 1 to 3). The local stability in this physical model relies on friction slowing the ball down after release from position 2 (otherwise it would be like a frictionless pendulum switching continuously from position 2 to position 3) and gravity. The concept of local stability can be seen by pushing the ball over the edge of the cup (Fig. 1.3c). Now it falls away from the edge at A or A′ towards another equilibrium point at B or B′. Finally, there is the possibility of an unstable equilibrium, the extreme of which can be visualized as a ball balanced on an infinitesimally small point (Fig. 1.3d). This equilibrium is mathematically stable but easily perturbed by a tiny displacement away from the equilibrium.

Real populations can be described in a similar way. For example, consider two interacting species (e.g. two competitors or a predator and a prey species); if the density of species B is plotted against the density of species A then perturbations of the densities of A or B will reveal the stability boundaries. In Fig. 1.4. reduction of the density of species A to y results in return to the initial (equilibrium) density x, whilst reduction to z pushes the system beyond the local stability boundary. This could be undertaken as a removal experiment in which the density of species A is reduced by different amounts and the return of it and other (possibly competing) species to an apparent equilibrium investigated, e.g. the removal of *Spartina patens* and other plant species from a marsh in North Carolina (Silander & Antonovics 1982). Investigation of stability in this way over the ecologically realistic range of densities will reveal the global stability of the system. Alternatively, one can look

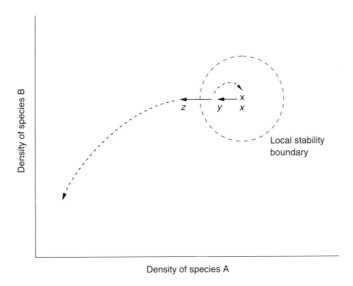

Fig. 1.4 Density of plant species A plotted against the density of plant species B and illustration of local stability by displacement from equilibrium (x) to position y or z.

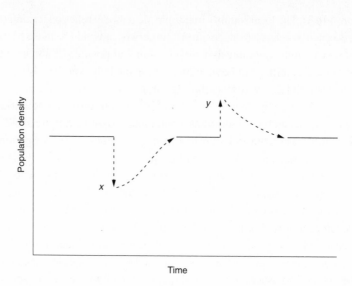

Fig. 1.5 Perturbation of time series away from equilibrium by reduction to density x or increase to density y. In both cases the population returns to the equilibrium showing it to be locally stable.

at the dynamics of a population over time (Fig. 1.5) and perturb it to different degrees at certain points in time and see whether it returns to the same (apparent) equilibrium. For example, in Fig. 1.5 increase above or decrease below the steady state reveals the population to be locally stable between the densities x and y. These perturbations are much easier to do in a computer than a field but the possibility does exist for such examinations of stability. Indeed, they may occur as a result of natural experiments, e.g. a drought year reducing the density of a plant population. A graphical description of stability is developed in Chapter 3 with analysis of local stability in Chapters 5 and 7.

1.7 Conclusion

Mathematical models in ecology can be categorized according to their simplicity, rationale and formulation. This introductory chapter presents a rather polarized view of these phenomena: simple versus complex, stochastic versus deterministic and so on. These are intellectual conveniences appropriate to an initial consideration. There are a whole series of intermediate models which can only be appreciated by further study. So read on!

CHAPTER 2

Probability of population extinction

Conservation biologists are addressing both the mechanisms underlying population extinctions and ways of assessing the likelihood of extinction (e.g. Soule 1987, Pimm *et al.* 1988, Foley 1994, Taylor 1995). A recurring problem in conservation biology is knowing how to use population abundance data and population dynamics theory to evaluate which populations of which species should be protected or managed. International conservation organizations, such as the International Union for Conservation of Nature and Natural Resources (IUCN), have become increasingly interested in the application of population biology theory to conservation problems, beginning with Mace and Lande (1991) and culminating in the new IUCN Red List categories approved in November 1994 (Species Survival Commission 1994). In this chapter we will examine how modelling techniques can be used to assess the likelihood that a population will become extinct over a given time period.

2.1 Defining the probability of population extinction

It is important to distinguish between two types of extinction. One type, a purely **stochastic** one, depends only on fluctuations in the population, perhaps caused by external events such as weather patterns (Fig. 2.1a). We assume that there is no change in the mean population size for this first type of extinction. The second type of extinction depends on the fact that the mean population size reduces predictably over time, sending the population towards extinction (Fig. 2.1b), i.e. a **deterministic** extinction process. This may be caused by a variety of factors, such as habitat loss or hunting. A combination of these two types of extinction can also be envisaged, i.e. a population with a declining mean size which also fluctuates about the mean (Fig. 2.1c).

Fluctuations in population size over time, such as those for the Bay checkerspot butterfly (*Euphydryas editha* ssp. *bayensis*, Fig. 2.2), may be large and unpredictable. In any year there is the possibility that a sufficiently large fluctuation will cause the population to become extinct. An estimate of the **probability of extinction** allows us to quantify that possibility. We can define probabilities of extinction for populations over given periods of time—a population might have a 1 in 10 chance of becoming extinct in any one year. We need to determine a probability of extinction because of the underlying stochasticity—we cannot be certain if extinction will occur in any one year. To begin with a density-independent deterministic model will be constructed (Section 2.2.1) which will shed light on deterministic extinction.

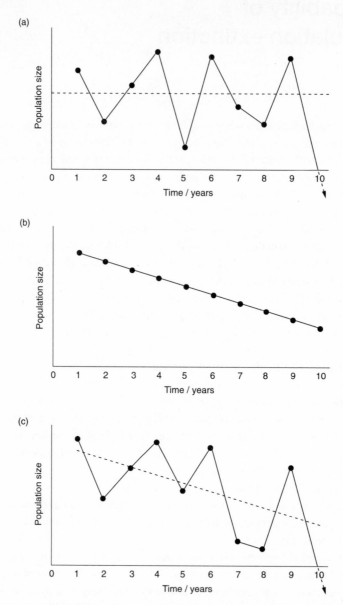

Fig. 2.1 Types of population extinction. (a) Extinction due to fluctuations around a constant mean population size (a stochastic process). (b) Extinction due to a decreasing mean population size (a deterministic process). (c) A combination of (a) and (b).

Population characteristics which lead to stochasticity will then be incorporated into this model (Section 2.2.3) and further refined by considering the effects of density dependence (Section 2.3).

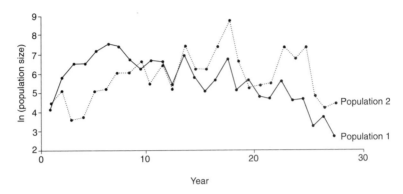

Fig. 2.2 Fluctuations in numbers of female Bay checkerspot butterfly (*Euphydryas editha* ssp. *bayensis*) over time at two locations in the San Francisco Bay area, California (data in Harrison *et al.* 1991).

2.2 A density-independent model for predicting probabilities of population extinction

2.2.1 Construction of a deterministic density-independent model

We will start by assuming that a population has a certain initial size at time 1. (Note that the population sizes need not be strictly interpreted as total population sizes but as densities based on a particular sampling area.) The individuals in this population are assumed to have fixed mean values of fecundity and survival at different ages. Fecundity and survival are therefore **parameters** of the model, i.e. variables which are held constant under particular circumstances, such as for certain species in a given habitat. A simple model species is one which breeds only once a year and then dies after reproduction, i.e. it has discrete (separate) generations. Certain insect species and annual plants fall into this category. We will begin by constructing a model for the cinnabar moth (*Tyria jacobaeae*) as many details of its life-history and population dynamics are well known (Dempster & Lakhani 1979, Crawley & Gillman 1989, van der Meijden *et al.* 1991). This is a tactical model (Chapter 1) which will be simplified wherever possible and from which a strategic model will emerge. Various methods of model simplification for cinnabar moth dynamics are considered in Gillman and Crawley (1990). It will be assumed throughout this chapter that the population is closed, i.e. there is no emigration or immigration. The life-cycle of the cinnabar moth and examples of its dynamics at three locations in Europe are given in Fig. 2.3.

There may be death at various stages in the life-cycle (Fig. 2.3aii), e.g. not all of the eggs will survive to become larvae nor all of the larvae to become pupae. In late spring the adults emerge and each female may lay up to 300 eggs (Dempster & Lakhani 1979). Although there is variation in fecundity between individuals, often

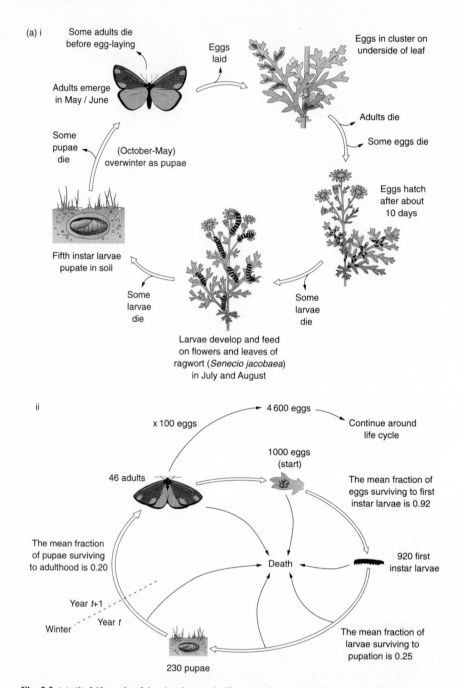

Fig. 2.3 (a) (i) Life-cycle of the cinnabar moth (*Tyria jacobaeae*) and (ii) example of increase in population size over one year with the density-independent model (see text). The values shown are the fraction of individuals surviving from one stage to the next and the average number of eggs produced per individual adult (see text). The larvae feed predominantly on ragwort (*Senecio jacobaea*).

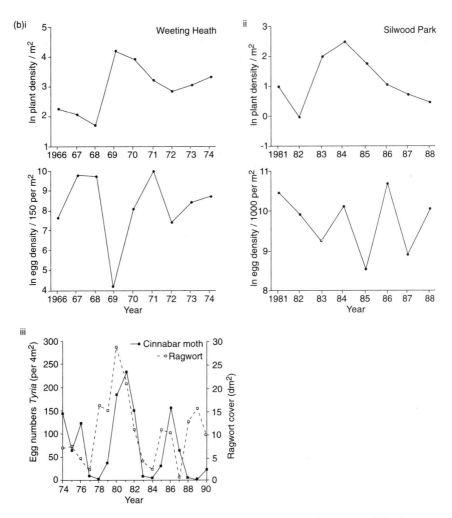

Fig. 2.3 (*continued*) (b) Population dynamics of the cinnabar moth and ragwort at (i) Weeting Heath, England (Dempster & Lakhani 1979); (ii) Silwood Park, England (Crawley & Gillman 1989) and (iii) Meijendel sand dunes, Holland (van der Meijden *et al.* 1991).

dependent upon larval food supply (which in turn affects pupal size and adult fecundity), in this first model we will ignore this variation and take an average of 200 eggs per female. This represents an average of 100 eggs per individual adult given a 1:1 sex ratio. The fraction of eggs surviving has been estimated as 0.92 (this and the following survival estimates are taken from Dempster & Lakhani 1979). We can now begin constructing a population model for the cinnabar moth. Assume the population begins with 1000 eggs (Fig. 2.3aii):

1000 eggs × 0.92 gives 920 first instar larvae.

From the 920 first instar larvae which begin feeding on ragwort (*Senecio jacobaea*) in June, it has been estimated that the fraction surviving until pupation will vary from 0.5 to less than 0.01. Although the actual survival rate depends, amongst other things, on larval food availability, we will again ignore the variation and assume that an average of 0.25 of first instar larvae survive until pupation:

920 first instar larvae × 0.25 gives 230 pupae.

Finally, between 0.02 and 0.45 of pupae survive to emerge as adults during May to June of the following year. Again we assume an average survival rate (0.20):

230 pupae at end of this year × 0.2 gives 46 adults next year.

Therefore, of the original 1000 eggs at the beginning of one year, only 46 have become adults in the following year. At each point in the life-cycle we have allocated a fixed survival value. We can combine all the survival values to give an (average) overall survival value from egg to adult:

0.92 (egg survival) × 0.25 (larval survival) × 0.20 (pupal survival)
= 0.046 (overall survival value).

Recalling that each adult 'produces' on average 100 eggs, we can now complete a simple population model, starting with 1000 eggs:

1000 eggs in year *t* results in 46 adults in year *t* + 1 which in turn produces 46 × 100 = 4600 eggs in the same year (*t* + 1).

More generally:

number of eggs in year *t* × 0.046 (overall survival) × 100 (fecundity)
= number of eggs in year *t* + 1.

To generalize further let the number of eggs in year *t* be E_t, the number of eggs in year *t* + 1 be E_{t+1} and *lambda,* λ, equal the survival values and fecundity multiplied together, i.e. 0.046 × 100 = 4.6. *Lambda,* λ, is often referred to in the ecological literature as the **finite rate of population increase** (or more strictly population *change* as populations can decrease in size; this value is referred to by May (1981) as the 'multiplicative growth factor per generation'). The population dynamics of the cinnabar moth (and other species with similar life-histories) can now be expressed as:

$$E_{t+1} = \lambda E_t \tag{2.1}$$

Equation 2.1 is an example of a first-order difference equation (Box 2.1). It is first order because it relates numbers or density at *t* + 1 to numbers or density one time step earlier (*t*). Note that λ will be the same if calculated from any corresponding life-history stage in successive years (e.g. adult to adult rather than egg to egg).

Rearranging Eqn 2.1 we can see that the finite rate of population increase can also be defined as the number of eggs produced in year $t + 1$ divided by the number of eggs in year t ($E_{t+1}/E_t = \lambda$), e.g. $4600/1000 = 4.6$. Thus 4.6 eggs in one year are produced for every egg in the previous year.

Box 2.1 Structure of difference equations

Difference equations are also known as recurrence relations. In ecological applications they often take the form of numbers (N) or biomass of a species at time $t + 1$ related to numbers or biomass at previous time t, e.g.

$$N_{t+1} = 4N_t$$

Therefore once N_1 is known, N_2, N_3 and so on can be determined. The numbers or biomass may also be related over space giving rise to an equation such as $N_{s+1} = 10 N_s$, i.e. the numbers at position $s + 1$ are 10 times the numbers at position s. These equations are first-order because they relate numbers at t or s to numbers one step on (or back). A second-order equation involves two steps, e.g. $N_{t+1} = 3N_t + 2N_{t-1}$ which could also be written as $N_{t+2} = 3N_{t+1} + 2N_t$.

If the survival and fecundity values remain constant, the population will increase by a multiple of 4.6 (λ) every year. It is generally true that a population composed of individuals with constant fecundity and survival values will have a constant rate of population change. If λ is less than 1 then deterministic extinction (Fig. 2.1b) will occur. There is also the possibility of no change in population size with time if the death rates are exactly balanced by the birth rates (i.e. $\lambda = 1$). The properties of density-independent models represented by Eqn 2.1 are explored in Questions 2.1 and 2.2.

Question 2.1

If a population starts with 100 eggs in year 1 (t), and has increased to 250 by year 2 ($t + 1$), what is the finite rate of increase assuming the density-independent model of Eqn 2.1? If this rate is constant, what will be the size of the population in years 3 and 4?

Question 2.2

Assume that there are three different moth populations, all of which begin with 10 eggs and follow the density-independent model of Eqn 2.1. The finite rates

Continued on page 18

Question 2.2 (*continued*)

of increase of these three populations are 2, 1 and 0.5, respectively. (a) Plot the change in population size over 4 years for each population. (b) How would you describe the change in population size of these three populations? (c) What is the value of λ required for the populations *not* to go extinct? (d) Is the model for population increase at $\lambda = 2$ realistic? (e) Finally, try replotting the data using a logarithmic scale (Box 2.2) for the population sizes. What advantage does the logarithmic plot have over the first graph?

Box 2.2 Logarithms and powers

Any number can be expressed as another number (the base) raised to a particular power, for example:

$100 = 10^2$ or 4.64^3

Powers can be added and subtracted when there is a common base, e.g. with base 10:

$10^3 = 10^{2+1} = 10^2 \times 10^1$

$10^4 = 10^{6-2} = 10^6/10^2$

In general, for a base x and powers a and b:

$x^{a+b} = x^a \times x^b$

$x^{a-b} = x^a/x^b$

Powers can include fractions. These are equivalent to combining square, cubic or higher roots and powers, e.g.

$2^{1/2} = \sqrt{2}$

$2^{1/3} = \sqrt[3]{2}$ (cube root of 2)

$2^{2/3} = (2^{1/3})^2$ or $2^{2/3} = \left(\sqrt[3]{2}\right)^2$

$2^{5/3} = (2^{1/3})^5 = \left(\sqrt[3]{2}\right)^5$

Note the special case of power 0; any number raised to power $0 = 1$.

Negative powers are equivalent to the reciprocal of a number raised to a positive power, for example, $2^{-3} = 1/2^3 = 1/8$.

A logarithm (abbreviated to log) is simply the power with a given base (which may be any number), for example, $100 = 10^2$ which is 10 to the power

Continued

Box 2.2 (*continued*)

2. 10 is the base and therefore 2 is the logarithm to the base 10 of 100, written as $\log_{10} 100 = 2$.

In general, if $a = x^m$ then $\log_x a = m$.

Just as powers can be manipulated so can logs. With $m = \log_x a$ and $n = \log_x b$:

$$\log_x(a \times b) = \log_x a + \log_x b = m + n$$

$$\log_x(a/b) = \log_x a - \log_x b = m - n$$

A useful result is that an expression of a^b can be converted into $\log_{10}$ (rather than $\log_a$) to give $b \log_{10} a$. In other words it is possible to convert between bases (see natural log examples below). Also ca^b can be converted to $\log_{10} c + b \log_{10} a$. For example, the species (*S*)–area (*A*) curve described by $S = cA^z$ can be converted to $\log_{10} S = \log_{10} c + z \log_{10} A$, enabling $\log_{10} c$ (the intercept on the vertical axis) and z (the gradient) to be found by linear regression of $\log_{10} S$ against $\log_{10} A$.

a^x is an example of an exponential function in which x is the exponent.

Plotting y against x for given values of a (≥ 0) produces a family of exponential curves (Fig. 1). These are often referred to as geometric change in y with changing x.

Several features of the curves in Fig. 1 are noteworthy. As x becomes a larger negative value so the value of y asymptotically approaches 0 when $a > 1$. In ecological modelling and mathematics generally one is frequently interested in the change in a dependent variable (e.g. probability of extinction) as an explanatory variable (e.g. population size) becomes very large or very small. A second feature of the curves in Fig. 1 is that when $x = 0$, $a^0 = 1$ and therefore all the curves in Fig. 1 pass through $y = 1$ when $x = 0$. As x becomes larger and

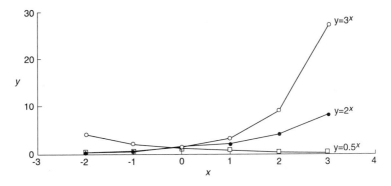

Fig. 1 Examples of exponential curves with $y = 0.5^x$, $y = 2^x$ and $y = 3^x$. When $y = 1^x$ the line is horizontal.

Continued on page 20

Box 2.2 (*continued*)

positive the values of *y* rapidly approach very large positive numbers (for $a > 1$ the same occurs if $a < 1$ and *x* becomes smaller and negative). A special case of these exponential curves is one where the gradient is equal to *x* at all points on the curve. Therefore, when $x = 0$, the gradient of the curve is equal to 1. Investigation of this special case led to the discovery of the irrational number e which has a value of 2.718 to four significant figures. e^x is known as the exponential function defined by $y = e^x$ in which the gradient at any value of *x* equals e^x. We can therefore have logs to the base e (known as natural logs and denoted by ln).

Conversion between natural logs and logs to the base 10 can be undertaken using the general rule for converting between bases, e.g.

$e^2 = 7.39$

$\ln(e^2) = 2$ (by definition)

To convert to $\log_{10}$:

$\log_{10}(e^2) = 2 \log_{10}(e) = 2 \times 0.4343 = 0.869$

$\operatorname{antilog}_{10}(0.869) = 7.39$, i.e. our starting point.

(A handy rule of thumb to get from ln to $\log_{10}$ of any number is to divide by 2.303, or multiply by 2.303 to get from $\log_{10}$ to ln.)

2.2.2 Estimating λ from field data

We will now consider how λ is estimated from field data on annual plants. Annuals germinate in autumn or spring, grow and reproduce within a year, and then die after flowering. This provides us with discrete generations, which can be modelled using first-order difference equations (Box 2.1). As our model is density-independent, it will be more accurate to consider populations at low densities (Question 2.2d). At higher densities, competition with neighbouring plants for resources may change rates of survival and fecundity, so that our assumption that these may be regarded as constant will be contravened.

Recalling the previous example of the cinnabar moth, λ needs to be estimated by censusing a given life-history stage over time. For an annual plant λ could be calculated by focusing on, e.g. seedlings which could be censused in consecutive years. λ would then be calculated as follows:

$$\lambda = \frac{\text{Seedlings censused during year } t + 1}{\text{Seedlings censused during year } t}$$

Table 2.1 Estimates of λ for three genotypes of an annual plant (oilseed rape) when introduced into new habitats in three successive years.

	Genotype 1	Genotype 2	Genotype3
Year 1	3.00	1.90	2.30
Year 2	0.15	0.04	0.10
Year 3	5.74	4.93	4.37

This seems very straightforward. However, if we are considering a population that is established in a habitat, then it may already have reached densities at which competition for resources is an important factor, so that the assumptions of density independence are not valid. λ will be most informative when a plant population is in the initial stages of the invasion of a habitat. As it is unlikely that an ecologist will be present at such a natural rare event, this could be part of the design of a field experiment. A known number of seeds would be sown into the habitat, and the number of seeds produced at the end of the growing season recorded. There may be some other factors to take into account, for example, seeds of many plants exhibit dormancy, which should be estimated and included in the calculation for λ. This also prohibits us from running the experiment in a habitat which may contain seeds of this plant species already.

Table 2.1 illustrates estimates of λ from field data for three genotypes of oilseed rape (*Brassica napus*), an annual plant (Crawley *et al.* 1993). In this case, the assumptions of the density-independent model were justified; the plant was being introduced in low densities into habitats in which it had not previously been grown. Interestingly, it can be seen that in this example when the same experiment was repeated in different years, the estimates of λ varied to a greater extent between years than between genotypes. Indeed, the final estimate of λ for any one genotype can fluctuate above or below 1 which is the deterministic extinction threshold.

Such variation, and the variation in survival and fecundity noted for the cinnabar moth, suggests that λ should not be regarded as a constant, but as a parameter that fluctuates within defined limits. The consequences of allowing λ to fluctuate are pursued in the next section.

2.2.3 *Introducing stochasticity into the model*

We are now ready to consider a model in which λ is not constant but is able to fluctuate. Let us further assume that the factors causing the fluctuations do not necessarily have the same effect each year. In fact we are going to assume that they are random in their operation; in other words there are good years and bad years for the populations and these are entirely unpredictable in their occurrence over time.

These unpredictable factors could affect λ in a variety of ways. For example, with the cinnabar moth, high June rainfall might reduce the survival rate of eggs and early instar larvae whilst unusually high winter temperatures might increase the survival of seeds of an annual plant. In order to explore the effect of such fluctuations (climatic or otherwise) let us examine a model population described by $N_{t+1} = \lambda N_t$ (Eqn 2.1 using N instead of E) where N is the population density of an annual species. We will assume that λ can take a range of values, each with a certain probability. λ is no longer described by its mean, but by a **probability density function** (pdf). In this case we will assume the discrete probability distribution illustrated in Fig. 2.4. A continuous pdf such as the normal distribution may be more appropriate but we will consider that at the end of this chapter. (Further details of pdfs are given in Box 6.1.)

The chosen pdf in Fig. 2.4 has only six values, each of which has the same probability (an example of a uniform distribution). A computer program may be written which takes a value of λ at random from this distribution and multiplies it by an initial population size at time 1 (N_1) to give the value of N_2. N_2 is then multiplied by a new value of λ, plucked again at random from the pdf, to give N_3 and so on. The six possible values of λ can be matched into three pairs: 1/10, 10; 1/2, 2; 3/4, 4/3. So, for example, the population has the same chance of being halved as being doubled. We might therefore expect the net change in population size to be zero, i.e. that the population will fluctuate around its original level. However, there is a chance of a number of bad years in a row, which might lead to extinction. For example, five very bad years in a row, starting from an initial population size of one, would give a population size of $1 \times (1/10)^5$. It is necessary to set an arbitrary extinction density greater than zero because the population will not reach zero given the assumptions of the model. In fact, with successive bad years the population moves asymptotically towards zero. Once the extinction density

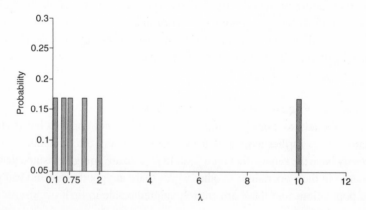

Fig. 2.4 An assumed probability density function for λ.

is set a proportion of the population following the dynamics described by $N_{t+1} = \lambda N_t$ and the pdf in Fig. 2.4 may become extinct.

We will now explore the effect of the assumed pdf on the population dynamics of the model population. Using initial population sizes of 1, 5 and 10, the model dynamics were simulated with λ following the pdf illustrated in Fig. 2.4. The simulation was repeated 10 times for each initial population size, and stopped after 20 generations (years) if the population had not become extinct. An arbitrary extinction density of 0.5 was assumed. The results of this exercise are illustrated in Table 2.2.

It can be seen that increasing the initial population size from 1 through to 10 decreased the probability of population extinction and increased the mean persistence time—an intuitively sensible outcome. It may seem surprising that with an initial population size of 1, any of these populations persisted at all. The fact that 3 out of 10 did persist is a consequence of the high variance in the distribution of λ (a phenomenon which you could explore in your own models).

We now come to an interesting predicament. λ was made to follow a uniform distribution, the arithmetic mean of which is 2.45 (check this for yourself). Despite this it is not predicted that the population would increase in size, if it persists. This is because we are predicting the mean effect of *multiplying* N_t by λ, not the mean effect of adding λ. In our example, the pdf was so simple we could see that, on average, the mean effect would be as if λ was 1 (indeed, we chose the values for that reason!). Thus, the net effect of multiplying N_t by the six λ values (1/10, 1/2, 3/4, 4/3, 2 and 10) was to multiply by 1, i.e. not to change the original value of N when averaged over many years. Rather than the arithmetic mean it is the **geometric mean** that is of relevance in this context. The geometric mean of a set of r numbers is found by multiplying them together and taking the rth root. In our example the geometric mean of 1 is found by multiplying the six values together and taking the sixth root. For more complicated pdfs, it can be easier to predict the outcome if the model is transformed into an additive rather than a multiplicative model. This can be achieved by taking logarithms (to a given base) of $N_{t+1} = \lambda N_t$, giving:

$$\log N_{t+1} = \log \lambda + \log N_t \qquad (2.2)$$

Table 2.2 Simulation results for a density-independent model $N_t = \lambda N_{t+1}$. Each simulation was for 20 generations and was replicated 10 times with λ following the pdf illustrated in Fig. 2.4.

Initial population size	Probability of extinction	Mean persistence time (generations)
1	0.7	8.8
5	0.6	12.4
10	0.3	16.7

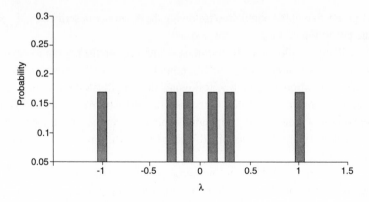

Fig. 2.5 pdf of $\log_{10}(\lambda)$ corresponding to pdf in Fig. 2.4.

If we look at the pdf of $\log_{10}\lambda$ we can see that the arithmetic mean is *zero* (Fig. 2.5, the bars are symmetrical about zero), and therefore on average (from Eqn 2.2), $\log N_{t+1} = \log N_t$. (A logarithm without a stated base should be assumed to be $\log_{10}$.) The geometric mean of λ is obtained by calculating the arithmetic mean of $\log \lambda$ and back transforming (taking the antilog). In this case the arithmetic mean of $\log \lambda$ is 0, which is equivalent to a geometric mean of 1. Hence, we can see that in cases where the expression is a multiplicative one, the geometric mean is more informative than the arithmetic mean.

In this section we have made important simplifications concerning the properties of populations, namely, that under fluctuating environmental conditions they may drift continually upwards or downwards (until extinction) in size over time and that under constant conditions they may increase or decrease geometrically. Such drifting and geometric change is unrealistic under most conditions. Real populations often seem to be limited in their size; some are more or less consistently abundant whilst others are more or less consistently rare. Populations that persist over long periods of time are presumed to be regulated in some way—by definition (as closed populations) they have not previously declined to extinction and, whilst they may occasionally reach very high densities, their numbers will then drop. The mechanism underlying this regulation is referred to as density-dependent change in survival, 'mortality' or fecundity (or, more briefly, **density dependence**). For example, as the density of organisms increases there is an increase in the fraction of individuals dying, i.e. mortality is no longer constant (for a particular age or stage of organism) but is determined by the density of the organisms. If we are to make our models more realistic then we must understand how density dependence affects population dynamics and incorporate it into these models. This is undertaken in the next section.

Question 2.3

How could the stochastic modelling exercise in this section be improved without altering the rules of the model?

Question 2.4

Do you expect that the results of the modelling exercise in Table 2.2 would have been the same if you had done it?

2.3 The effect of density dependence on the probability of population extinction

2.3.1 Density-dependent population regulation

With the models of insect and annual plant population dynamics we might expect both competition and attack by natural enemies to act as density-dependent factors and therefore affect population change. For example, as the population density increases, resources become depleted and intraspecific competition may become increasingly important (e.g. Hassell 1975, Watkinson 1980). If natural enemies are attracted to areas of high prey density, then individuals may have a higher probability of attack by predators at high densities than they do at low densities. It is these assumptions of factors altering with population density which we will now incorporate into the population dynamics model previously summarized by Eqn 2.1. For illustration we will consider a model insect population, although the resultant models and techniques are applicable to other organisms with discrete generations.

The essence of all density-dependent mechanisms is that, as the population density increases, there is an alteration in fecundity or the fraction of individuals surviving and therefore a change in λ, the finite rate of population increase. In Fig. 2.6(a) a set of lines have been plotted, describing the ways in which the fraction of individuals surviving (s) may alter with population density (N). Examples of effects of density on survival and fecundity are shown in Fig. 2.6(b). The horizontal line in Fig. 2.6(a) indicates density-independent survival which is already part of our model. Note that with density-independent survival the numbers dying increase with density, but the fraction dying remains constant. We will first consider the simplest type of density-dependent response given in Fig. 2.6(a), the linear reduction in s.

The linear reduction in s with increasing N can be expressed as the equation of a straight line (Box 2.3):

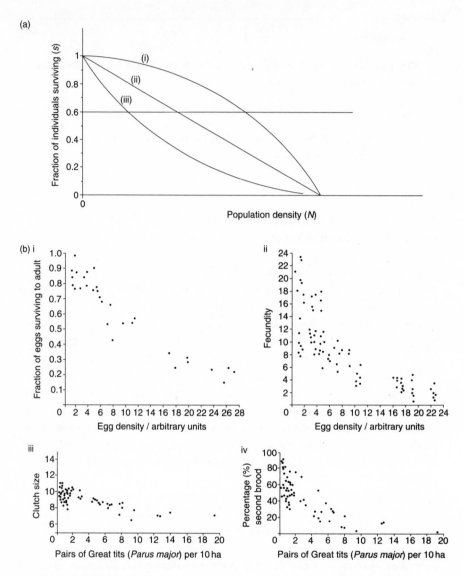

Fig. 2.6 (a) Hypothetical relationships between fraction of individuals surviving (*s*) and population density (*N*). (i) scramble competition, (ii) intermediate competition, (iii) contest competition. (b) Examples of density dependence: (i) reduction of fraction of *Drosophila melanogaster* eggs surviving to adulthood with increasing egg density, (ii) reduction of *Drosophila* fecundity with increasing egg density (i and ii from Prout & McChesney 1985), (iii) reduction in clutch size and (iv) reduction in percentage of second brood of Great tits (*Parus major*) with increasing density of nesting pairs (data of Kluiver in Hutchinson 1978).

$$s = c - mN \qquad (2.3)$$

where *c* is the intercept on the *s*-axis (*c* must lie between 0 and + 1) and *m* is the

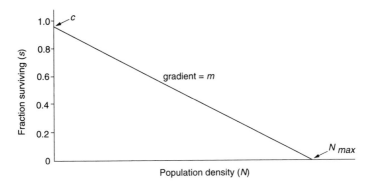

Fig. 2.7 Linear density dependence of fraction surviving, s, with density, N, with gradient, m, intercept, c and maximum density, N_{max}, shown.

magnitude of the gradient of the line. Thus, we assume at the lowest density that s is close to c, i.e. a fraction c individuals survive in the population and with any increase in density s declines linearly according to the gradient m. This relationship is shown in Fig. 2.7. The linear reduction in s with N assumes that there is a density, N_{max}, at which $s = 0$. In other words there is an upper (N_{max}) and lower boundary (0 or arbitrary approximation) of possible extinction. If a population goes above N_{max} in any one year it will become extinct. If it goes below the arbitrary extinction density it will also become extinct. These two boundaries are important to bear in mind when the probability of population extinction is considered.

Box 2.3 Linear and quadratic functions

A polynomial is a function of the form:

$$f(x) = a + bx + cx^2 + dx^3 + \ldots + jx^n$$

where n is a positive integer (including 1).

The order or degree of a polynomial is the highest power of x in the polynomial. In this text we will be concerned with polynomials of order 1 or 2, described below.

Linear function (polynomial of degree 1)

$$f(x) = a + bx \qquad a \text{ can equal } 0$$

which may form one side of the familiar equation of a straight line $y = a + bx$ where a is the intercept on the y-axis and b is the magnitude of the gradient. a and b may be positive or negative.

Continued on page 28

Box 2.3 (*continued*)

Quadratic function (polynomial of degree 2)

   $f(x) = a + bx + cx^2$ a and/or b can equal 0

   This can be one side of a quadratic equation $y = a + bx + cx^2$. Plotting y against x gives a parabolic curve (Fig. 1), the shape and location of which depend on the values of a, b and c.

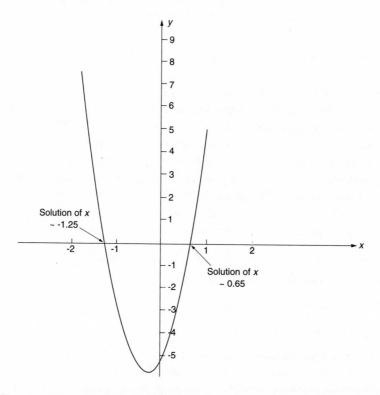

Fig. 1 Parabolic curve produced by the quadratic equation, $y = 6x^2 + 4x - 5$. The approximate solutions of the equation are given by the intersection with the x-axis where $y = 0$.

   If y is known in the above equations then there are methods of finding the value(s) of x, i.e. solving the equations.

Solution of linear equation
Any linear equation, for example:

Continued

Box 2.3 (*continued*)

$$3 = -2 + 4x$$

can be rearranged and x (= 5/4) found directly.

Solution of quadratic equation
Consider the quadratic equation:

$$3 = 6x^2 + 4x - 2$$

This is first put into standard format by setting the LHS equal to 0 (in this case by subtracting 3 from both sides):

$$0 = 6x^2 + 4x - 5$$

In a quadratic equation there will be two values of x, i.e. two solutions or roots. The simplest general method of solution is to use the equation:

$$x = \frac{-b \pm \sqrt{b^2 - 4ac}}{2a}$$

where a, b and c are coefficients in the general quadratic equation $ax^2 + bx + c = 0$. So, for the above example we have:

$$x = \left[-4 \pm \sqrt{4^2 - 4 \times 6 \times -5} \right] / 2.6$$

$$x = \left[-4 \pm \sqrt{136} \right] / 12$$

$$x = 7.662/12 = 0.639 \ \text{ or } \ x = -15.662/12 = -1.305$$

A rough check of these solutions is given by the graph of $y = 6x^2 + 4x - 5$ (Fig. 1, note the values of x at $y = 0$). The solutions of a quadratic equation can be positive, negative or complex.

From Fig. 2.7 m can be expressed as c/N_{max}. Substituting this version of m into Eqn 2.3 gives an alternative way of expressing s:

$$s = c - (c/N_{max})N$$

$$s = c\,(1 - N/N_{max}) \tag{2.4}$$

In our cinnabar moth example, there were, on average, 92 first instar larvae emerging from every 100 eggs laid. The larvae then fed on ragwort during June to August with an average of 0.25 of the number of first instar larvae surviving to pupation. If we assume that the food plant is a limited resource, then larvae may compete with each other (intraspecifically) for food. The effect of competition on individual survival would be expected to become more intense as larval density (L) increased. We can therefore imagine density-dependent survival during the larval stage. In our new model the density-independent average of 0.25 can be taken as the value of c, i.e. when the effects of density are negligible. Let us assume that as the density of larvae increases so the fraction of larvae surviving falls linearly according to Eqn 2.4 (replacing N by L). Note that the larval density is determined before any larvae die and that L and L_{max} refer to larvae in the same year (t).

We can now construct a new model, incorporating density dependence, to replace the previous density-independent model (see p.16). The larval survival rate is now dependent upon the initial number of larvae following Eqn 2.4:

E_t eggs × 0.92 (egg survival) × $[0.25(1-L/L_{max})]$ (larval survival) × 0.20 (pupal survival) × 100 (average fecundity) = number of eggs in the following year (E_{t+1})

Note that if $L = 0$ we revert to the density-independent equation. Combining the constant survival and fecundity values into the single value, λ (= 0.92 × 0.25 (c) × 0.20 × 100), leaves the following equation incorporating density dependence:

$$E_{t+1} = E_t \lambda \left(1 - \frac{L}{L_{max}}\right) \tag{2.5}$$

This can be made neater by noting that L is the number of larvae which hatch in year t and therefore is related to the number of eggs in year t:

$$L = 0.92E_t \quad \text{and} \tag{2.6a}$$

$$L_{max} = 0.92E_{max} \tag{2.6b}$$

Substituting Eqn 2.6(a) and (b) into Eqn 2.5 yields:

$$E_{t+1} = E_t \lambda \left(1 - \frac{E_t}{E_{max}}\right) \tag{2.7}$$

In other words, we can now predict the number of eggs in year $t + 1$ based on the number of eggs in the previous year (t). This is a major step forward. Equation 2.7 is an example of the **discrete logistic equation** which is written for numbers of any stage (N) as:

$$N_{t+1} = \lambda N_t \left(1 - \frac{N_t}{K}\right) \tag{2.8}$$

where K is the maximum number of that stage; K is often referred to as the **carrying**

capacity, defined as the maximum density of a species that can be supported in a habitat. Equation 2.8 can also be expressed as:

$$N_{t+1} = \lambda N_t (1 - \alpha N_t)$$

replacing $1/K$ by α. Berryman (1992) and Elliott (1994) review the use of these and similar equations and May *et al.* (1974) and May (1981) discuss the density-dependent terms. The discrete logistic equation is a strategic model which can be applied to a wide range of examples, but which, as witnessed above, has its foundation in the real life-histories of individuals of particular species.

We must now consider some of the properties of Eqn 2.8 before evaluating what effect density dependence has on our estimate of the probability of extinction. (You will need to keep remembering that the equivalent density-independent equation (Eqn 2.1) was only able to increase or decrease geometrically or stay at an unstable equilibrium, see Fig. A1, p. 182.) If we multiply out the brackets on the RHS of Eqn 2.8 we discover an important attribute of this density-dependent equation:

$$N_{t+1} = \lambda N_t - \frac{\lambda N_t^{\,2}}{K} \tag{2.9}$$

Equation 2.9, the expanded form of Eqn 2.8, demonstrates clearly that the equation is **non-linear** (a quadratic equation because of the N_t^2, Box 2.3). It is this non-linear component that gives the equation some fascinating properties (including the possible production of **chaotic dynamics**) which we will explore in the next section.

2.3.2 Exploration of the dynamics produced by the density-dependent model

Equation 2.8 is an example of a **first-order non-linear difference equation**. These equations form the backbone of many models of population dynamics and we will accordingly spend some time considering their behaviour here. In exploring their behaviour we have two options. First, given initial conditions, e.g. an initial value for egg numbers, and parameter values for K and λ, we can generate a series of egg values in much the same way as we did for the density-independent version of the equation. Such a procedure rapidly becomes extremely tedious by hand and therefore it is normal to write a computer program to perform the task and plot out the results. This is a **simulation** approach to the exploratory process—it will show us what the equation (model) can do but not necessarily tell us much about why it does it. If we want to know why, then we have to carry out some form of mathematical analysis, referred to as the **analytical** approach. Some simple analytical techniques are detailed after the simulations.

Values of N generated from simulations using Eqn 2.8 with $K = 100$ and $\lambda = 2$, 3.1, 3.5 and 4.0, respectively, are displayed in Fig. 2.8, starting with 10 eggs. Figure 2.8(a) shows a flow diagram of the sequence of calculations in the simulation. (Such simulations can be undertaken in a variety of software packages.) This is an

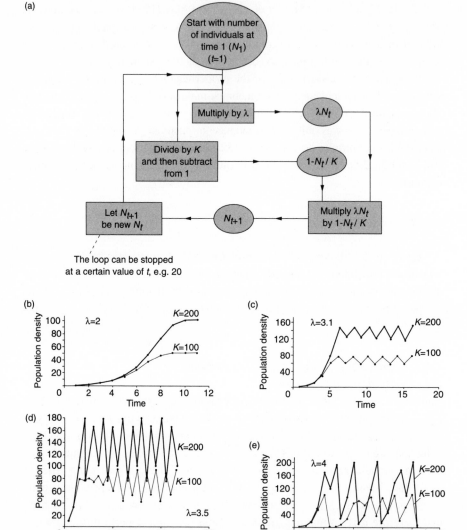

Fig. 2.8 (a) Flow diagram of sequence of calculations showing how to generate successive values of N_t based on the discrete logistic Eqn 2.8. Change in density (N_t) with time generated from Eqn 2.8 with (b) $K = 100$ and 200 and $\lambda = 2$. (c) $K = 100$ and 200, $\lambda = 3.1$. (d) $K = 100$ and 200, $\lambda = 3.5$ (e) $K = 100$ and 200, $\lambda = 4.0$.

iterative process, as discussed for the density-independent model, in which we generate an answer for N_{t+1} and then use it as the new N_t and so on. You should check the first few iteration values given in Fig. 2.8(b).

The dynamics of the model population over time at different values of λ can be summarized as follows. At $\lambda = 2$ the population approaches an equilibrium value of 50 at which it remains (Fig. 2.8b), i.e. this appears to be its stable equilibrium value. At $\lambda = 3.1$ (Fig. 2.8c) the population oscillates between two densities; this is referred to as two-point **limit cycles**. At $\lambda = 3.5$ (Fig. 2.8d), four-point limit cycles are produced whilst at $\lambda = 4.0$ (Fig. 2.8e) the initially regular cycles break up, so that the population fluctuates, apparently unpredictably, between a series of densities. This is referred to as chaotic dynamics. In this equation the values of K do not affect the dynamics and only contribute to the size of the equilibrium (e.g. Fig. 2.8b,c).

To be certain of the stability of the equilibrium with $\lambda = 2$ we need to perturb the equilibrium (Chapter 1). This can be achieved by using different initial conditions and checking the steady state. This would show that the equilibrium of 50 is indeed stable—in fact it is globally stable for ecologically realistic values. The two-point limit cycle is also stable—for a given value of λ the population will always settle to fluctuate between the same two densities. The chaotic dynamics do *not* have this property—the particular sequence of values is dependent upon initial conditions, although the size of the fluctuations will be determined by the model. Box 2.4 gives a short description of chaos in ecology.

Box 2.4 Chaos in ecology

The possibility of chaotic dynamics means that if 'random' or unpredictable dynamics are recorded in field populations this does not necessarily imply that the underlying mechanisms are random. Some or all of the 'randomness' could be produced by predictable deterministic processes expressed as chaos. Thus, if population change is described by the discrete logistic equation (2.8) each population size at $t + 1(N_{t+1})$ is given by a particular value of N_t. However, the sequence of values is unpredictable when λ is sufficiently high (Fig. 2.8e).

The problem is to distinguish chaotic dynamics from true randomness (stochasticity). The first study to look at chaos in laboratory and field populations was by Hassell *et al.* (1976). They used the technique of assuming an underlying deterministic model (described by the equation $N_{t+1} = \lambda N_t(1 + aN_t)^{-b}$, see Section 2.3.4) and ascertaining the values of λ, a and b for different populations of insect. They were then able to compare these values with those known to produce

Continued on page 34

Box 2.4 (*continued*)

limit cycles and chaos (Fig. 1). So Hassell *et al.* were really testing whether the model that is fitted to the data has parameter values which would give chaos. The parameter values of b and λ for each species (Fig. 1) were superimposed on the regions of different dynamic behaviour predicted by the model, e.g. stable equilibrium, limit cycles and chaos.

Only one species was predicted to show chaotic dynamics and one to show stable limit cycles whilst the others were in the stable equilibrium region. It is worth noting that the apparently chaotic population was a *laboratory* population of blowflies set up by Nicholson in 1954. The debate on the importance of chaos rumbles on as new analytical techniques are used. For example, a study by Ellner and Turchin (1995) using three analyses, suggests that Nicholson's blowflies are on the edge of chaos and no populations (either in the laboratory or the field) are chaotic for all analyses.

In summary, chaos is a fascinating phenomenon which appears to be relevant to a wide range of disciplines from fluid dynamics to economics. In ecology its appearance from very simple population models cautions us against assuming

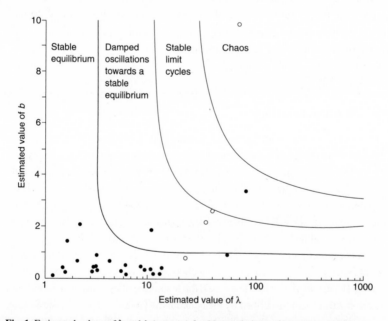

Fig. 1 Estimated values of λ and b (see text) for 28 populations of insect, overlaid on regions of different dynamic behaviour (open circles are data for laboratory populations; closed circles are from analyses of field data), from Hassell *et al.* (1976).

Continued

Box 2.4 (*continued*)

that all fluctuations are caused by stochastic events. The implications for population regulation are that density dependence is only stabilizing under certain conditions. If density dependence is coupled with high rates of population increase then chaos may result, which destabilizes the population.

The local equilibrium value in Fig. 2.8(b) can also be found analytically. This is achieved by realizing that at equilibrium $N_{t+1} = N_t$; thus there is no change in the population density over time. If we make the LHS of Eqn 2.8 equal to N_t then:

$$N_t = \lambda N_t \left(1 - \frac{N_t}{K}\right)$$

Dividing both sides by N_t and rearranging, we have:

$$N^* = K\left(1 - \frac{1}{\lambda}\right) \tag{2.10}$$

The N_t value in the equation is the equilibrium value, and so this has been denoted by N^*. This can be checked by substituting values for K (= 100) and λ (= 2), to give $N^* = 50$, which agrees with the result obtained by simulation (Fig. 2.8b).

A useful graphical method for analysis of first-order non-linear difference equations is to plot N_{t+1} against N_t (a Ricker–Moran plot). This provides information on the stability of the system described by the non-linear difference equation and the likelihood of limit cycles or chaos. Take the discrete logistic equation 2.8. First plot out the curve described by the equation and then draw the line of $N_{t+1} = N_t$ (the line of unity). The point of intersection of the straight line and the curve tells us the value of N at which the equilibrium occurs. This is a graphical method of solving Eqn 2.10. To follow the dynamics on a Ricker–Moran plot we begin at an initial value of N_t, say 20 (Fig. 2.9). The next value of N (N_{t+1}) is read off the curve and, using the line of unity ($N_{t+1} = N_t$), N_{t+1} is changed to N_t and the process repeated. This eventually gives us the two-point limit cycles seen in Fig. 2.9 for a value of $\lambda = 3.1$ and $K = 100$.

It is fairly easy to determine analytically the value of λ at which limit cycles begin, because it is known that when the gradient of the curve at $N_{t+1} = N_t$ is equal to -1, limit cycles begin (May & Oster 1976). For example, for the discrete logistic, we begin by differentiating Eqn 2.8 (if you are unfamiliar with differentiation read Box 3.1) with respect to N:

$$N_{t+1} = \lambda N_t - \lambda N_t^2/K \qquad \text{(Expanded form of Eqn 2.8)}$$

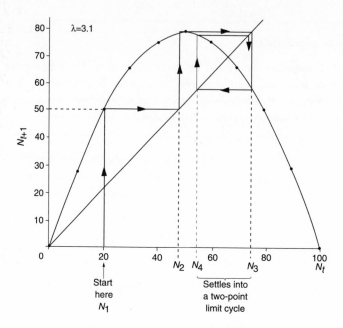

Fig. 2.9 Iterations of N_t on a Ricker–Moran plot using Eqn 2.8 with $\lambda = 3.1$ and $K = 100$.

$$\frac{dN_{t+1}}{dN_t} = \lambda - \frac{2\lambda N_t}{K}$$

$$= \lambda\left(1 - \frac{2N_t}{K}\right)$$

Set dN_{t+1}/dN_t equal to -1:

$$-1 = \lambda\left(1 - \frac{2N_t}{K}\right) \qquad (2.11)$$

Now solve this at equilibrium N^*.

Substitute $N^* = K\left(1 - (1/\lambda)\right)$, i.e. Eqn 2.10 in Eqn 2.11:

$$-1 = \lambda\left(1 - \frac{2K\left(1 - \frac{1}{\lambda}\right)}{K}\right)$$

Cancel the Ks to give:

$$-1 = \lambda - 2\lambda + 2\lambda/\lambda$$

Cancel the λs and rearrange:

$$\lambda = 3.$$

Therefore the stable equilibrium ends and limit cycles begin at $\lambda = 3$ which agrees with the simulations in Fig. 2.8. Note again that the carrying capacity K is not relevant to the dynamics in this equation.

Question 2.5

In Fig. 2.10 there are two time series (*ai* and *aii*) and two Ricker–Moran plots (*bi* and *bii*). Time series *ai* is from a randomly fluctuating population and *aii* from a population with chaotic dynamics produced by a non-linear difference equation. Which Ricker–Moran plot corresponds to which time series and why?

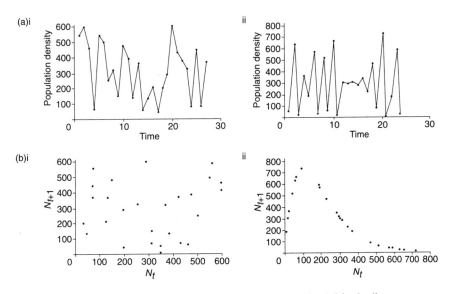

Fig. 2.10 (a) Two time series. (b) Two Ricker–Moran plots. See Question 2.5 for details.

2.3.3 Determination of the probability of extinction

We shall now introduce a stochastic element into the density–dependent model (using Eqn 2.8) in the same way that we did with the density-independent model. λ is made to follow the same pdf as illustrated previously (Fig. 2.4), and the model is run for 20 generations with initial population sizes of 1, 5 and 10. The results of 10 such simulations are illustrated in Table 2.3.

Table 2.3 Simulation results for a density-dependent model (Eqn 2.8) with a pdf for λ as defined in Fig. 2.4.

Initial population size	Probability of extinction	Mean persistence time (generations)
1	1.0	4.3
5	1.0	6.8
10	1.0	10.3

When the result in Table 2.3 is compared with the density-independent model (Table 2.2), we see that the probability of extinction has increased and mean persistence time decreased. Indeed the probability of extinction is the maximum of 1 in each case. At first this seems a little counter-intuitive, as we have introduced density dependence into our model, which should have a stabilizing influence. However, it is the result of the particular assumptions in Eqn 2.8. We have assumed that at the maximum density (K) and above, there is zero survival. When close to K, the population may still leap above it in one generation, with the result that it becomes extinct in the next generation. So what we have done is to introduce an upper boundary as well as a lower boundary at which the population goes extinct— hence the higher probabilities of extinction. With the parameter estimates we have chosen, it is extremely likely that the population will reach one or other boundary within 20 generations.

To remove the upper boundary at K we need to assume a *non*-linear relationship between density and survival, e.g. asymptoting towards 0 at the highest densities (see the example in Fig. 2.6a). This is explored in Section 2.3.4.

2.3.4 The consequences of non-linear density dependence

One of the easiest ways of incorporating non-linear density dependence is to use an exponential function (Box 2.2) so that the fraction of individuals surviving (s) declines with density (N) according to a negative exponential function:

$$s = e^{-aN} \tag{2.12}$$

Compare Eqn 2.12 with Eqn 2.3. The change in s with increasing values of N for two different values of a (0.1 and 0.01) is shown in Fig. 2.11.

The parameter a can be thought of as denoting the strength of density dependence. At any given value of N, the fraction surviving will decrease as a increases. When $N = 0$, s will be equal to 1, regardless of the value of a, i.e. the model is designed so that at very low densities s tends towards a maximum value. Conversely, at very high densities, the value of s tends towards 0 (but does not reach it as in the linear density dependence example).

We can now write a new equation representing the population dynamics of a species with non-linear density dependence:

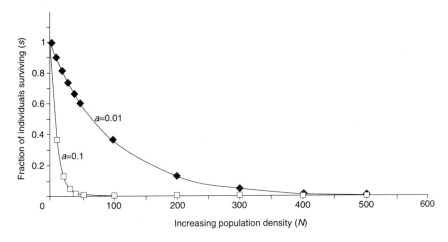

Fig. 2.11 Exponential decrease in fraction of individuals surviving, s, with increasing density, N at two values of a using Eqn 2.12.

$$N_{t+1} = \lambda N_t e^{-aN_t} \tag{2.13}$$

Equation 2.13 is another first-order non-linear difference equation, known as the Ricker equation (Ricker 1954). Once again, simulations to determine the probability of extinction can be generated to compare with our two previous models (Table 2.4).

Both the probabilities of extinction and the mean persistence times are comparable to the density-independent model (Table 2.2). This is because the model has no upper extinction boundary, like the first density-dependent model, but has the same lower extinction boundary as both previous models and therefore a similar probability of population extinction to the density-independent model.

The examples in Sections 2.3.3 and 2.3.4 illustrate the importance of considering the mechanisms underlying density dependence, as this has profound implications for the shape of the density-dependent function and therefore the stability of the model. The different curves in Fig. 2.6(a) can be considered to reflect different types of intraspecific competition. An important distinction is between scramble

Table 2.4 Results from 10 simulations of a density-dependent model (Eqn 2.13) with $a = 0.001$ and with a pdf for λ as defined in Fig. 2.4.

Initial population size	Probability of extinction	Mean persistence time (generations)
2	0.8	8.8
5	0.6	12.3
10	0.4	15.9

(curve i) and contest (curve iii) competition (Hassell 1975). In pure scramble competition, resources are divided equally amongst competing individuals. The consequence of this is that above a certain density, the mean resource per individual is too low to allow survival, and therefore s plummets to 0. Thus the upper boundary in the discrete logistic model (curve ii) may be taken to represent perfect scramble competition, although the overall shape of scramble is better represented by curve (i). The other extreme is represented by contest competition, in which the superior competitors monopolise the resources. Consequently, a certain number of individuals always survive, even at high densities (curve iii). In this case, s would approach 0 at high densities, but never reach it, in agreement with the Ricker equation (Eqn 2.13). The results of the extinction simulations in Tables 2.3 and 2.4 suggest that scramble competition is relatively destabilizing compared to contest competition.

It is therefore possible to envisage a range of types of density dependence produced by intraspecific competition, from pure scramble through to pure contest. The model of Hassell (1975) described by the equation $N_{t+1} = \lambda N_t (1 + aN_t)^{-b}$ provides parameters a and b which can describe the change from contest to scramble. The model is worth reviewing not only in the context of scramble and contest competition but also as an example of the development of a model. This model is closely related to the simpler fisheries model of Beverton and Holt (1957) $N_{t+1} = \lambda N_t (1 + aN_t)^{-1}$ but derived from $N_s = (1/\alpha)N_t^{(1-b)}$ where N_s = density of survivors and N_t is the original density. This was used by Morris (1959) and Varley and Gradwell (1960, 1963) to detect density dependence by taking logs and rearranging to give:

$$\log N_s = -\log \alpha + (1 - b) \log N_t$$

$$\log \frac{N_t}{N_s} = \log \alpha + b \log N_t$$

They used linear regression of $\log(N_t/N_s)$ (the k-value, killing power) against $\log N_t$ to estimate the strength of the density dependence b (see Section 2.4 for discussion of this regression method). However, a linear regression is often inappropriate (Fig. 2.12a). Frequently, below a threshold density ($\log N_c$) the R-value changes very little with density.

To overcome this the Hassell (1975) model was derived in which a gives the threshold density at which density dependence occurs ($a = 1/N_c$) and b is the strength of density dependence. The effect of these parameters can be seen by plotting $\log (N_t/N_s)$ against $\log(N_t)$ (Fig. 2.12b,c), which gives the required non-linear regression rather than the linear regression above. To get to this regression take N_{t+1} as λN_s and divide both sides of the Hassell equation by λ to give $N_s = N_t(1 + aN_t)^{-b}$. Then take logs and rearrange:

$$\log N_s = \log N_t - b \log(1 + aN_t)$$

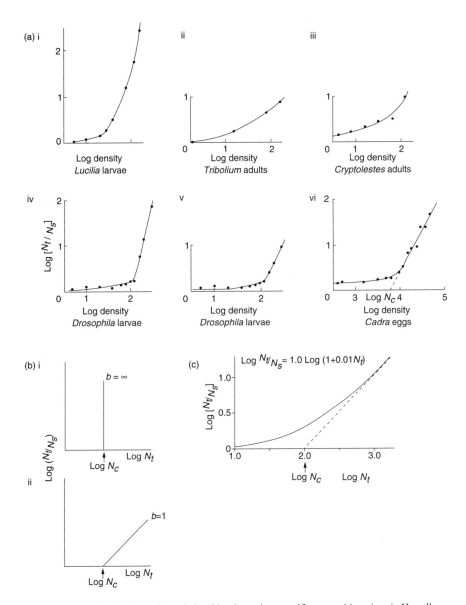

Fig. 2.12 Some density-dependent relationships due to intraspecific competition given in Hassell (1975). In each case mortality, or reduced natality, is expressed as a k-value (log N_t/N_s) and plotted against the log initial density (log N_t). Figs (a)i to (a)v, after Varley *et al.* (1973); Fig. (a)vi, after Rogers (1970). All curves fitted by eye. (a) Mortality of *Lucilia cuprina* (Meig.) between larval and adult stages (data from Nicholson 1954). (b) Reduced fecundity of *Tribolium castaneum* (Herbst) (data from Birch *et al.* 1951). The density-dependent relationships arising from two extremes of intraspecific competition: (i) scramble; (ii) contest. Axes as in Fig. 2.13(a). (c) Reduced fecundity of *Cryptoplestes*; a density-dependent relationship from the Hassell equation where $a = 0.01$ and $b = 1.0$.

$$\log\left(\frac{N_t}{N_s}\right) = b \log(1 + aN_t)$$

Note that the RHS of the equation is non-linear (e.g., Fig. 2.12c).

The parameter b serves to distinguish between scramble and contest competition above the threshold a. In the case of scramble a very small increase in density above the threshold leads to a dramatic increase in K-value; i.e. b is very large (Fig. 2.12 bi). In contrast pure contest competition will leave a constant number of survivors and therefore as density increases b will equal 1 (Fig. 2.12bii). The model has been subsequently used to test for chaos (Hassell *et al.* 1976; see Box 2.4), used as the basis of an interspecific competition model (Chapter 5, Hassell & Comins 1976) and developed to analyse competition within plant species (Chapter 4, Watkinson 1980).

2.4 Estimation of probability of extinction from field data

Until now we have guessed the parameter values for the density dependence in our models, whilst the estimates of λ have come from combining mean field estimated values of egg, larval and pupal survival with fecundity and sampling from an assumed pdf. One method of estimating density dependence that can also be used to estimate values for λ is to use time series data, e.g. a series of annual censuses, which can be in one or more sites. As discussed for the density-independent model, it is also possible (and often preferable) to set up field or microcosm experiments in which population densities and habitat variables (such as food resources) are manipulated in such a way that density dependence parameter values can be estimated. For example, we could assume a particular form of density-dependent model (e.g. negative exponential) in the examples of Fig. 2.6(b) and estimate parameter values by regression. Parameter estimation linked to field manipulations are, despite widespread interest, quite rare (see Gillman *et al.* 1993, described in Chapter 4).

We will now consider a simple time series method for estimating the density dependence parameter value, with the bonus of an estimate for λ that can be compared with the estimate that combines survival and fecundity.

Consider the density-dependent model described by the Ricker equation (2.13):

$$N_{t+1} = N_t \lambda\, e^{-aNt}$$

A convenient way of estimating the two parameter values of λ and a is to use linear regression techniques, proceeding as follows. Divide both sides by N_t and take the natural logarithms:

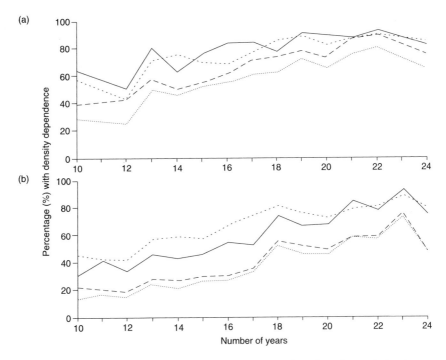

Fig. 2.13 Percentage of time-series with significant (5%) density dependence detected using three methods: (———) Bulmer's; (- - - -) Ricker's; (– –) Pollard *et al.*; (....) all methods; in (a) aphids and (b) moths.

$$\ln\left(\frac{N_{t+1}}{N_t}\right) = \ln(\lambda) - aN_t \qquad (2.14)$$

Now, if we plot $\ln(N_{t+1}/N_t)$ against N_t we can estimate $\ln(\lambda)$ from the intercept and a from the gradient. Although this is convenient we need to be cautious as this method can also detect density dependence from a time series of random numbers, i.e. it may often detect density dependence when it is not there. Woiwod and Hanski (1992) used it as one of several measures of density dependence for 94 species of aphid and 263 species of moth collected in the Rothampsted Insect Survey. This comprehensive analysis showed detection of density dependence increased with census duration to between 60 and 80% with the Ricker equation being one of two methods giving the highest detection rate (Fig. 2.13).

We are now ready to deal with some real data and attempt to estimate the values of a and λ using Eqn 2.14. Of course, in choosing particular data sets to analyse, we are assuming that our model is a reasonable description of the dynamics of that population.

2.4.1 Estimation of λ and density dependence for the cinnabar moth

These data are taken from the same site where values of survival and fecundity were used to estimate λ (Dempster & Lakhani 1979; Section 2.2.1). The first problem we encounter is the choice of a suitable measurement of density. The larvae of the cinnabar moth feed on ragwort, the density of which fluctuates considerably from year to year (Fig. 2.3b). It is the intensity of larval competition for food which is the mechanism behind the density dependence in this system. The density-dependent survival of larvae therefore depends upon larval density per unit of plant resource, rather than on population density per unit area. The appropriate model therefore becomes:

$$N_{t+1} = N_t \lambda \, e^{-\frac{aN_t}{P_t}} \qquad\qquad (2.15)$$

with P_t representing the biomass of plant in year t. Thus the density is not simply the density of eggs or larvae but, more realistically, the density per unit amount of plant material (N_t/P_t). From the Weeting data set we can now regress $\ln(N_{t+1}/N_t)$ against N_t/P_t (Fig. 2.14, using egg densities) to find the values for λ and a in this model:

$\ln(\lambda) = 1.89, \qquad \lambda = 6.63$

(compare this with the average value for λ of 4.6 in Section 2.2.1)

$a = 0.325$

$r^2 = 0.654$

The value of r^2 tells us the proportion of the total variance that is explained by

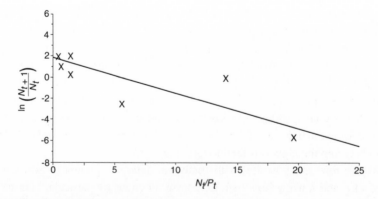

Fig. 2.14 Regression of ln N_{t+1}/N_t against N_t/P_t, using the linearized version of Eqn 2.15, to estimate ln(λ), the intercept on the vertical axis and a, the gradient.

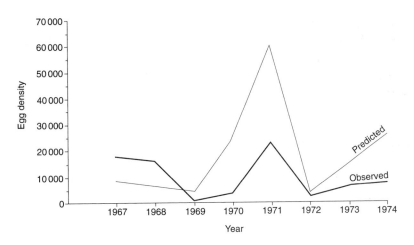

Fig. 2.15 Predicted and observed dynamics of the cinnabar moth, *Tyria jacobaeae*.

the regression model; in this case, over half. We can use our parameter estimates of λ and *a* and the initial population density to produce a population trajectory and compare this with the observed population dynamics (Fig. 2.15). Although the observed and expected values look similar this is not too surprising as the model parameters were taken from the same data set. A more rigorous test of the model would be to use parameter estimates from one data set to generate time series for other data sets, (Chapter 1). The comparison of observed and predicted time series would then act as an independent test of the model.

2.4.2 Estimation of probability of extinction for cinnabar moth

In Section 2.4.1 we dealt with parameter estimates from entirely deterministic models, in which every value of N_t has a corresponding value of N_{t+1} given by the particular equation we are using. However, it is clear from the examples of real field data above that such models are not sufficient to explain all the variation in N_{t+1} there is a scatter of points around the regression line quantified by the unexplained variance. There is more than one contribution to this variance. If a population under observation genuinely followed a deterministic trajectory, sampling error alone would cause the measurement of population density to vary around predicted values. This will be especially apparent if populations are clumped or individuals are difficult to sample, e.g. soil-living insects. There will also be variation from individual to individual, so that small populations in particular will have a range of average survival and fecundity (and therefore λ) values. The latter is referred to as **demographic stochasticity**. Finally, there may be variation in λ or in the strength of density dependence from one year to the next because of variation in environmental factors. This is referred to as **environmental stochasticity**. Our models may be adapted to take account of these contributions to the unexplained variance. We began doing this in Section 2.2.3 for the density-independent model when λ was varied according to a given pdf. What is needed now is a way of incorporating the unexplained variation estimated from the field into the model. We will then be in a position to consider its effect on the probability of extinction.

In this case, we will assume that λ takes a particular mean value for a site or habitat which itself does not vary over time. This mean value is multiplied by a value taken randomly from a continuous pdf—the pdf does not vary with time but the value selected will vary from year to year. We will denote the pdf by ε and define it as being normally distributed. So our density-dependent example (Eqn 2.15) now becomes:

$$N_{t+1} = N_t \varepsilon_t \lambda N_t\, e^{-a\frac{N_t}{P_t}} \tag{2.16}$$

You may recall from Eqn 2.2 that the effect of varying λ on N_t was illustrated by taking natural logs and in Eqn 2.13 the values of λ and a were estimated by dividing Eqn 2.12 by E_t and taking natural logs. In the same way we will illustrate the effect of ε:

$$\ln \frac{N_{t+1}}{N_t} = \ln \varepsilon + \ln \lambda - a\frac{N_t}{P_t} \tag{2.17}$$

For no overall effect of $\ln \varepsilon$ on $\ln \lambda$ (and therefore $\ln(N_{t+1}/N_t)$) we require that the mean of $\ln \varepsilon = 0$. This corresponds to the mean of ε being 1. Moreover, with a normal distribution of mean 0 and variance v, $\ln \varepsilon$ is equivalent to the error term in the linear regression of $\ln(N_{t+1}/N_t)$ against N_t/P_t.

We now see an interesting possibility. If we estimate the unexplained variation around the regression line we can quantify the pdf ε_t, and then begin to estimate the probability of extinction for different populations by sampling from it. The stochastic element is included as a continuous normal distribution in this case, rather than the discrete distribution of set values for λ that we have used in previous simulations. Simulations involving sampling from the estimated variance

Table 2.5 Simulation results for the cinnabar moth population model with $a = 0.325$ and $\lambda = 6.63$, incorporating a stochastic element $\ln \varepsilon$ which has a normal distribution with mean 0 and variance v (1.89).

Initial population size	Probability of extinction	Mean time to extinction (years)
5	1.0	6.2
10	1.0	7.2
50	0.9	7.9

around the regression in the Weeting Heath cinnabar moth data set are given in Table 2.5.

The values of a, λ and v used in Table 2.5 result in the prediction that even in areas of relatively high initial population densities (e.g. 50), there is a probability that the population will become extinct. This is a consequence of the strong density-dependent mortality experienced by the larvae of this moth, in competing for food, and of the high year-to-year variation represented by a v value of 1.89. McCarthy (1996) gives an example of similar methods involving density dependence applied to a vertebrate species (the red kangaroo).

2.4.3 Analytical technique for estimating probability of extinction applied to various taxa

Foley (1994) has applied analytical techniques to the problem of extinction in discrete time. He used the natural log transformed density-independent population model $\ln N_{t+1} = \ln \lambda + \ln N_t$ (Eqn 2.2). He assumed that $\ln \lambda$ was normally distributed with a variance v_r, that the population started at size N_0 and proceeded on a random walk with a maximum value of k and a minimum value of 0, i.e. extinction. The mean time to extinction, T_e, for a population starting at N_0 with no population growth (i.e. $\ln \lambda$ has a mean of 0) is then given as:

$$T_e = \frac{2\ln(N_0)}{v_r}\left[\ln(k) - \frac{\ln(N_0)}{2}\right] \tag{2.18}$$

This therefore employs an analytical method, closely related to the simulation in Table 2.5. We can use Eqn 2.18 to calculate values for T_e for the cinnabar moth and compare them with the values in Table 2.5. The results are given in Table 2.6 for three values of N_0.

The results of mean time to extinction using the analytical technique (Table 2.6) are quite similar to the simulation results (Table 2.5). Both predict low mean times to extinction (less than 10 years) for all three population sizes. The major difference in the models is the incorporation of density dependence in the simulation—although the analytical model is density-independent the inclusion of

Table 2.6 The mean time to extinction for the cinnabar moth using Eqn 2.18. Based on census data from Weeting Heath (Dempster & Lakhani 1979). v_r is 8.633 and $\ln(k)$ is 9.909.

Initial population size (natural log of size)	Mean time to extinction (years)
5 (1.609)	3.4
10 (2.303)	4.7
50 (3.912)	7.2

a ceiling of k provides some element of density dependence because populations cannot drift continually higher in density.

The cinnabar moth example is an interesting one in the context of population extinction. It appears from the simulations and the analytical methods of Eqn 2.18 that its probability of extinction is high. Yet it is a common and widespread species. The solution to this paradox is that we need to look beyond local population extinction to the **metapopulation** level. A metapopulation system is effectively a population of local populations each of which can go extinct (with a certain probability) but can also be formed by colonization. Therefore the births (colonization) and deaths (extinction) of the local populations in the metapopulation mirror the births and deaths of individuals in a local population. As will be seen in Chapter 6, metapopulation dynamics are dependent upon the relative values of local population extinction and colonization. For many species it is the metapopulation level that provides the most sensible estimates of species extinction. Indeed the methods described by Foley have now been applied in a metapopulation context (Hanski *et al.* 1996).

Foley applied the methods of Eqn 2.18 (and more complex methods, e.g. including population growth) to a range of animal species including the Bay checkerspot (see Fig. 2.2). Harrison *et al.* 1991 also analysed the probability of extinction for the two butterfly populations in Fig. 2.2, examining the effect of density dependence and removal of individuals for sampling. The estimates of T_e for a variety of taxa are summarized in Table 2.7.

Table 2.7 Estimates of mean time to extinction T_e using Eqn 2.17 for one butterfly and three mammal species from Foley (1994) and a plant species (*Orchis morio*), Gillman and Silvertown (1997). Four of the five estimates come from more than one population.

Species	Expected time to extinction (years)
Bay checkerspot butterfly (*Euphydryas editha* ssp. *bayensis*; 2 populations)	54–69
Wolf (*Canis lupus*; 5 populations)	19–237
Grizzly bear (*Ursus arctos horribilis*; 1 population)	1488
Mountain lion (*Felis concolor*; 6 populations)	1378–6107
Green-winged orchid (*Orchis morio*; 8 samples)	18–78

It would be unwise to interpret these estimates of T_e too strictly. They are based on a very simple model with a series of assumptions. Perhaps their most useful function is to provide an estimate of the *relative* likelihood of extinction, as a contribution to the IUCN Red List process, allowing species to be prioritized for conservation. The new IUCN criteria use the following probability of extinction levels for species with adequate data sets (Species Survival Commission 1994):

Critically endangered: 0.5 (50%) within 10 years or three generations, whichever is longer.

Endangered: 0.2 within 20 years or five generations, whichever is longer.

Vulnerable: 0.1 within 100 years.

To use the IUCN criteria with Eqn 2.18 we need to convert from probability of extinction to time to extinction. This can be done if extinction is viewed as a Poisson process (following Foley 1994; see Box 6.1) and so T_e is approximated as:

$$T_e = \frac{-t}{\ln(1-P)}$$

where P is the probability of extinction and t is the number of years over which extinction is measured. If average generation time is longer than 3–4 years the value of T_e needs to be increased because t is effectively increased. In terms of T_e the IUCN thresholds for short generation species (< 4 years) are therefore:

Critically endangered: 14.4 years.

Endangered: 89.6 years.

Vulnerable: 949.1 years.

i.e. approximately one order of magnitude difference in T_e between critically endangered and endangered and between endangered and vulnerable.

There are also examples in the literature of models of extinction which have used continuous time stochastic methods (e.g. Renshaw 1991, Goodman 1987). These models all present similar results to those in Tables 2.2–2.7, i.e. they predict the time to extinction for a population with known initial size and variance in abundance. Continuous time models are introduced in Chapter 3.

2.5 Conclusions

The probability of extinction analyses in this chapter reflect many of the tensions between the requirements of field-based ecologists and the activities of theoreticians producing models. Whilst the models are fraught with problems of inadequate sets of data and oversimplistic assumptions, they are valuable because they allow an objective prioritization of threats to the world's flora and fauna. Furthermore, the requirement for long runs of high-quality data can help to direct strategies for data collection in the field and therefore bring modellers and field ecologists closer together.

CHAPTER 3

Looking for cycles: the dynamics of predators and their prey

I agree with him (Lotka) in his conclusion that these studies and these methods of research deserve to receive greater attention from scholars, and should give rise to important applications. [Vito Volterra, 1927.]

A fascinating aspect of the dynamics of predators and their prey is their propensity to **cycle in abundance**. The phenomenon is found in both invertebrates and vertebrates (examples are given in Fig. 3.1). Not all predators and prey cycle, and some species do it in one part of their range but not in others. In this chapter we will use models to help us understand the dynamics of cycling in predators and prey and and why some species may cycle and others do not.

In Chapter 2 we assumed that our model populations had discrete generations or, in the case of populations with overlapping generations, that reproduction was discrete. This is obviously inappropriate for many species. Human populations, for example, do not synchronize their reproduction! In these cases we should think of reproduction as a continuous process. This means that difference equations are no longer appropriate descriptions of population dynamics. We require a different model framework, in which time is continuous, to represent the changes in population size. This new framework is provided by differential equations. We will begin our search for cycles in continuous time. Sections 3.1 and 3.2 set the background for discussion of the dynamics of the prey. In Section 3.3 an artificial predator is added and in Section 3.4 predators and prey are modelled together in continuous time. Finally in Section 3.5 we return to discrete time to overcome some problems raised by the continuous time approach.

3.1 An equation for density-independent population growth in continuous time

In Chapter 2 we began by considering a simple equation in discrete time to represent density-independent population change. This was $E_{t+1} = \lambda E_t$ (Eqn 2.1). If reproduction is continuous then the difference between t and $t + 1$ is vanishingly small and therefore change is instantaneous and described by differential equations. In the discrete time model it was found that population change was geometric in form (Fig. 3.2a). Now let us assume that we are dealing with a continuously reproducing population. For a description of continuous geometric population change (also called exponential population change) the separate points in Fig. 3.2(a) are replaced by a smooth curve (Fig. 3.2b).

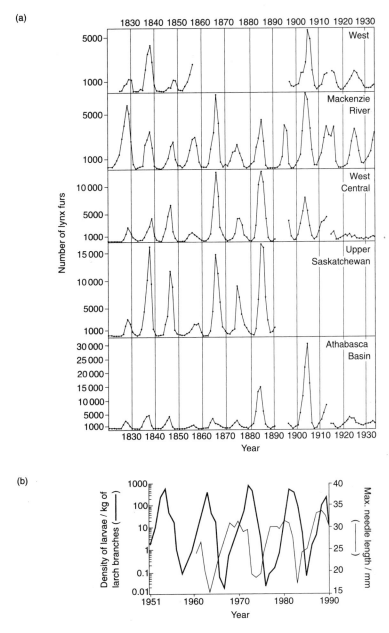

Fig. 3.1 Examples of cycles of abundance in predators. (a) Cycles in the number of lynx fur returns of the Hudson's Bay Company, from 1821 to 1934 (Elton & Nicholson 1942). (b) Cycles of abundance of the larch bud-moth and larch (Baltensweiler 1993).

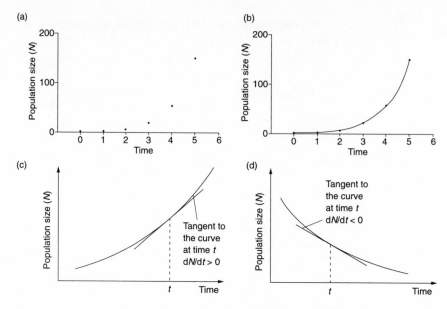

Fig. 3.2 (a) Geometric increase in discrete time. (b) Smoothed version of Fig. 3.2(a) in continuous time. Illustration of (c) positive and (d) negative rates of continuous population change.

If a population is changing exponentially then the curve can be described by the equation:

$$N_t = e^t \tag{3.1}$$

where N_t means the population size (or density) at time t. (For a discussion of the exponential function see Box 2.2.) At each point on the smooth exponential curve it is possible to determine the rate of population change by differentiation (Box 3.1) which is equivalent to the gradient of the tangent at that point, i.e. the population change is given by the derivative of N_t with respect to t, dN/dt. When dN/dt is positive the population is increasing over time (Fig. 3.2c); when dN/dt is negative the population is decreasing and when dN/dt equals zero there is no change in population size. The special property of the exponential curve is that $dN/dt = e^t$, i.e. the value of dN/dt is the same as the population size (N_t).

Box 3.1 Differentiation and differential equations

Differentiation of ax^n
Differentiation provides a way of finding the gradient of a curve at any point

Continued

Box 3.1 (*continued*)

on that curve and therefore of determining the rate of change of one variable in response to change in another. This is equivalent to determining the tangent of the curve at that point. Tangents are linear rates of change at one point on a non-linear curve (Fig. 3.2c and d). Drawing the tangent is only an approximate way of finding the gradient on the curve at a particular point. Differentiation produces a precise value.

Consider the function $y = ax^n$. The gradient at any point, written as dy/dx (described in speech as 'd y by d x') is referred to as the derivative of the function or the derivative of y with respect to x. The derivative of the function ax^n is:

$$\frac{dy}{dx} = anx^{n-1}; \text{ note that } \frac{dy}{dx} \text{ could be written } \frac{d(ax^n)}{dx}$$

For example, if $y = 3x^2$ then $dy/dx = 3.2x^1 = 6x$. Thus when $x = 2$, the gradient $= 12$. $dy/dx = 6x$ is an example of a differential equation.

x can also be raised to different powers:

$$y = 5x^4 + 4x^3 + 3x^2$$

the derivative of which is:

$$\frac{dy}{dx} = 5 \times 4x^3 + 3 \times 4x^2 + 3 \times 2x^1$$

$$= 20x^3 + 12x^2 + 6x$$

Therefore, if a function contains parts which are added or subtracted, the derivative is found by determining the derivatives of the parts. This is not the case when parts are multiplied (a product) or divided (a quotient) as shown in the following sections.

Differentiation of a product
Consider the function

$$y = e^x (3x^2 + 2x)$$

The derivative is found by taking the first part (e^x) and multiplying by the derivative of the second ($3x^2 + 2x$) and then adding the derivative of the first part multiplied by the second part, recalling that the derivative (gradient) of e^x is e^x (Box 2.2):

$$= e^x (6x + 2) + e^x (3x^2 + 2x)$$

$$= e^x (3x^2 + 8x + 2)$$

Continued on page 54

Box 3.1 (*continued*)

More generally, if $y = uv$ where both u and v are functions of x then

$$\frac{\mathrm{d}y}{\mathrm{d}x} = u\frac{\mathrm{d}v}{\mathrm{d}x} + v\frac{\mathrm{d}u}{\mathrm{d}x}$$

This can be extended to functions containing more than two parts multiplied together.

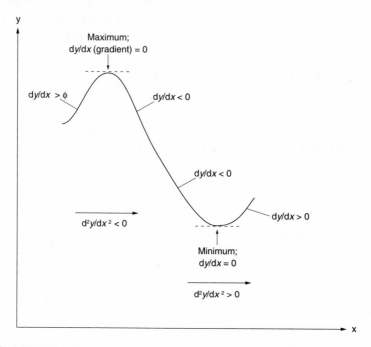

Fig. 1 Illustration of maximum and minimum values of a function $y = f(x)$.

Differentiation of a quotient
When one function of x (say u) is divided by another function of x (say v), i.e. $y = u/v$ the derivative is found by using the following formula:

$$\frac{\mathrm{d}y}{\mathrm{d}x} = \frac{v\dfrac{\mathrm{d}u}{\mathrm{d}x} - u\dfrac{\mathrm{d}v}{\mathrm{d}x}}{v^2}$$

Second-, third- and higher-order derivatives maybe found by continuing to differentiate. For example:

Continued

Box 3.1 (*continued*)

$$y = 5x^4 + 4x^3 + 3x^2$$

$$\frac{dy}{dx} = 20x^3 + 12x^2 + 6x$$

The RHS can be differentiated again to give the second-order derivative, written as d^2y/dx^2:

$$\frac{d^2y}{dx^2} = 60x^2 + 24x$$

The order of the derivative indicates the order of the differential equation. A useful application of second-order derivatives is in the determination of maximum and minimum values of a function (Fig. 1). Both have a gradient of zero (and therefore $dy/dx = 0$). To distinguish between them, the second-order derivative is examined. If it is positive then the value is a minimum (gradient goes from negative to positive) and therefore gradient is increasing with x, hence the positive value of d^2y/dx^2. The opposite applies to a maximum value.

Equation 3.1 assumes that the population size is 1 when $t = 0$. If it is not then we need to multiply by an initial value of N_0 (the population size at time $t = 0$) to give:

$$N_t = N_0\, e^t \tag{3.2}$$

The gradient of $N_t = N_0\, e^t$ at any point in time, t, is given by the value of dN/dt which is $N_0\, e^t$, i.e. it is again equal to the value of N_t. Therefore:

$$\frac{dN}{dt} = N_t \tag{3.3}$$

Equation 3.3 is a differential equation which describes a population changing exponentially with time. It states that the rate of population change at any point in time dN/dt is equal to the population size (N_t) at that point in time. This is obviously rather limited in its usefulness. It would be more helpful to express the rate of population increase as a multiple of the existing population size. To achieve this we can replace Eqn 3.2. by Eqn 3.4:

$$N_t = N_0\, e^{rt} \tag{3.4}$$

The rate of population change is again found by differentiating the equation with respect to t to give $dN/dt = rN_0\, e^{rt}$.

By substituting $N_t = N_0 e^{rt}$ (Eqn 3.4) into $dN/dt = rN_0 e^{rt}$, we see that

$$\frac{dN}{dt} = rN_t \tag{3.5}$$

This is a more useful equation than Eqn 3.3 because we now have the r term on the RHS. Therefore, the rate of population change at time t is equal to the population size at that time (N_t) multiplied by r which is referred to as the **intrinsic rate of increase** or **innate capacity for increase**. Strictly we should reserve the terms intrinsic or innate for the maximum value of r (r_m) which occurs under optimum conditions of temperature, light, food supply and so on. The *actual* instantaneous rate of increase, r, will always be lower than r_m and may vary from year to year or even day to day for a particular species. (r or r_m is also called the Malthusian parameter.) Equation 3.5 is an important equation, much used in ecological applications.

For a closed population, from and into which no migration occurs, Eqn 3.5 can be rewritten so that the parameter r is replaced by $b - d$:

$$\frac{dN}{dt} = (b-d)N_t \tag{3.6}$$

where b is the instantaneous birth rate (per individual) and d is the instantaneous death rate (per individual).

To illustrate the estimation of r we will use the example of population change in the United States from 1790 to 1910. Although these data were presented by Pearl and Reed (1920) to illustrate a different point, it is interesting to use them to contrast with their analysis, which is considered in the next section (3.2). To estimate the parameter r, we linearize Eqn 3.4 by taking the natural log (ln):

$$\ln(N_t) = \ln(N_0) + rt$$

This produces a straight-line equation relating $\ln(N_t)$ to t, with a gradient of r and intercept of $\ln(N_0)$ (Fig. 3.3a). r can then be estimated by linear regression, giving a value of 0.027. The linear fit is apparently very good, explaining 99.5% of the variance. However, the pattern of the residuals around the regression suggests that extrapolation of the linear model beyond 1910 may not be appropriate (Fig. 3.3b). If the linear model was appropriate we would expect an even scatter of points around the line (and even if it was appropriate we may still not wish to extrapolate beyond 1910).

In this example the value of r is estimated over a period of time when high levels of immigration were occurring in the United States and therefore is likely to be higher than r estimated for a closed population in which no immigration occurs.

To close this section we consider the relationship of the differential equation $dN/dt = rN_t$ (Eqn 3.5) to the difference equation $N_{t+1} = \lambda N_t$ (Eqn 2.1 replacing E by N). Both of these equations describe geometric or exponential population change,

(a)

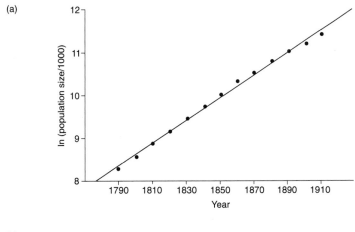

(b)

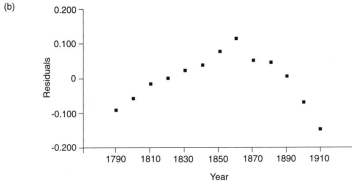

Fig. 3.3 (a) Growth of the human population in the USA from 1790 to 1910 (data in Pearl & Reed 1920). Data plotted as ln (population size ÷ 1000) against year (t). For example, in 1910 the population was estimated as 91 970 000. (b) Residuals of the linear regression in (a).

the first in continuous time and the second in discrete time. The rate of population change is given by r and λ respectively. But what is the relationship between these two parameters? Consider values of population density at two consecutive points in time. The differential equation 3.5 is derived from Eqn 3.4, $N_t = N_0 e^{rt}$. Therefore, with $N_0 = 1$, at $t = 1$, $N_1 = e^r$ and at $t = 2$, $N_2 = e^{2r}$. (Remember that this continuous time model can have values between $t = 1$ and $t = 2$, e.g. at $t = 1.5$, unlike the discrete time model.) Dividing N at time 2 by N at time 1 gives $e^{2r}/e^r = e^r$ (noting that $e^{2r} = e^r \times e^r$ and that any value of N_0 could have been used—it would cancel out here). Indeed, the continuously growing population increases by e^r between all consecutive integer time values. By comparison, the difference equation describes a change between consecutive integer time values as $N_{t+1}/N_t = \lambda$. Thus, λ is seen to be equal to e^r, or $\ln \lambda = r$.

Question 3.1

What are the conditions for increase or decrease in population size in Eqn 3.6?

3.2 Density-dependent population growth: the logistic equation

Both $N_{t+1} = \lambda N_t$ and $dN/dt = rN_t$ are density-independent equations because the growth parameters λ and r are unaffected by density. We can incorporate the ecological realism of density dependence into the differential equation for population change in the same way that we did for the difference equation model in Chapter 2. Consider the density dependence acting as a multiplier on r. (Look back at Chapter 2 if you need to be reminded of how mathematical statements of density dependence can be translated into ecological mechanisms such as intraspecific competition.) The simplest of several possible forms for the density dependence is a linear reduction in r with density (we used a linear reduction in λ in Chapter 2). This is achieved by multiplying r by $1 - N_t/K$ so that when $N_t = 0$, r is multiplied by 1 (and therefore not changed) but is steadily reduced as N increases until a value K is reached when $1 - N_t/K = 0$. K is again referred to as the carrying capacity, which is particular to a given habitat and may change with time. When $N_t = K$ there is no change in population size because $dN/dt = 0$, i.e. $N_t = K$ is an equilibrium point. When $N_t > K$, $1 - N_t/K$ becomes negative, therefore dN/dt is negative and population size decreases.

Multiplying r in the density-independent equation 3.5 by $1 - N_t/K$ gives the equation:

$$\frac{dN}{dt} = rN_t\left(1 - \frac{N_t}{K}\right) \tag{3.7}$$

Equation 3.7 is known as the **logistic equation** and it provides a simple description of the growth of some populations. Populations with dynamics described by this equation reach an equilibrium density of K without oscillations (Fig. 3.4a). We can confirm that K is the equilibrium size by rearranging the logistic equation at $dN/dt = 0$. At $dN/dt = 0$ the RHS of Eqn 3.7 will equal 0 which means that either $rN_t = 0$ or $1 - N_t/K = 0$. The first of these is a trivial solution: if $rN_t = 0$ then either the population size, N_t, is 0 or $r = 0$ (the latter tells us that the population is not growing, but does not give the equilibrium value). If $1 - N_t/K = 0$ then $N_t = K$, i.e. the population size at equilibrium is equal to K. The stability of the equilibrium is explored further in Section 3.3.

The logistic equation was first used by Verhulst (1838) to describe the growth of human populations and was used independently by Pearl and Reed (1920) to

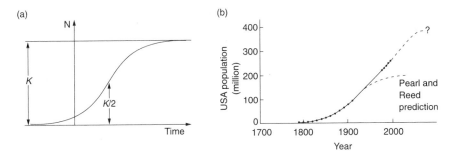

Fig. 3.4 (a) Shape of the logistic curve and relationship to parameters, reprinted from original in Pearl and Reed (1920). (b) Logistic growth of human population of the United States from Pearl and Reed (1920) and subsequent data.

describe human population growth in the United States (Fig. 3.4b). Hutchinson (1978) and Kingsland (1985) provide fascinating insights into the history of this work. You should compare Eqn 3.7 with Eqn 2.8 which is an example of the discrete logistic. Note that $1 - N_t/K$ can also be written as $(K - N_t)/K$.

It was clear from the residuals in Fig. 3.3b that the change in ln (population size) with time is non-linear. Pearl and Reed wanted to find a general mathematical description of population growth which starts as exponential growth, but thereafter in their words 'must develop a point of inflection, and from that point on present a concave face to the x axis, and finally become asymptotic, the asymptote representing the maximum number of people which can be supported on the given fixed area.' For this purpose they chose the following equation which produces a sigmoidal (s-shaped) growth curve as in Fig. 3.4b:

$$N = \frac{b\,e^{at}}{1 + c\,e^{at}}$$

where a, b and c are parameters for particular populations under certain environmental conditions.

The derivative dN/dt of their equation is:

$$\frac{dN}{dt} = aN\frac{1 - cN}{b}$$

which can be compared with Eqn 3.7 (see Question 3.2). Pearl and Reed fitted their equation to the population data in Fig. 3.3a. They calculated values of a, b and c by assuming that the population curve passed through three of the data points, thereby providing three simultaneous equations (Box 3.2) which could be solved to give values for a, b and c. This gave values of $r = 0.0313$ and $K = 197\,000\,000$. The value of r is close to our linear estimate of 0.027 (Fig. 3.3a). We have the benefit of knowing that, seventy years on, the current population of the United States exceeds Pearl and Reed's estimate of K by about 60 million (Fig. 3.4b, the 1992 estimate was 255 million). Furthermore, it is apparent that the population increase in the last twenty years is still high and positive so we are not yet close to K (in fact r is 0.01 for the period 1974–1992). The lack of agreement between Pearl and Reed's estimated K and current population size may be due to a variety of causes. For example, the logistic curve is symmetrical so that the population increase before the point of inflection must equal the population increase after (Fig. 3.4a). This may be unrealistic for many population growth curves. Related to this point, Pearl and Reed's estimates were based on data prior to the point of inflection. Indeed, it may be that the point of inflection has still not been reached (Fig. 3.4b). Human carrying capacity is also dependent on changing technology, so that predicted values of carrying capacity based on agricultural productivity and health care in the nineteenth century would inevitably be much lower than those predicted following the green, biotechnology and medical revolutions in the twentieth century. Nevertheless, Pearl and Reed believed that their model provided a useful and simple description of the mechanisms underlying population growth (human or otherwise):

> In a new and thinly populated country the population already existing there, being impressed with the boundless opportunities, tends to reproduce freely, to urge friends to come from older countries, and by the example of their well-being, actual or potential, to induce strangers to immigrate. As the population becomes more dense and passes into a phase where the still unutilized potentialities of subsistence, measured in terms of population, are measurably smaller than those which have already been utilized, all of these forces tending to increase the population will become reduced. Pearl and Reed (1920, p. 287).

The logistic curve has also been fitted to the growth of non-human populations with varying success. The best examples are cultures of algae, bacteria, insects and yeast (Fig. 3.5) where the simple assumptions of the logistic equation are most appropriate. The examples in Fig. 3.5(b) and (c) illustrate the sensitivity of r and K to genotype and different environmental conditions.

In the next section we will use the logistic equation as a description of population growth of a prey species subjected to harvesting by a human predator.

Fig. 3.5 (*opposite*) Examples of applications of the logistic growth curve. (a) Growth of yeast *Saccharomyces cerevisiae* populations in culture. From Allee *et al.* (1949) reprinted in Maynard Smith (1974). (b) Growth of friut fly *Drosophila melanogaster* populations. (i) wild type in a pint bottle; (ii) heterozygous or homozygous individuals for five recessives including vestigial wing in a pint bottle; (iii) wild type in a half-pint bottle. (c) Growth of water flea *Moina macrocopa* populations at three different temperatures. (b and c reprinted from Hutchinson 1978.)

(a)

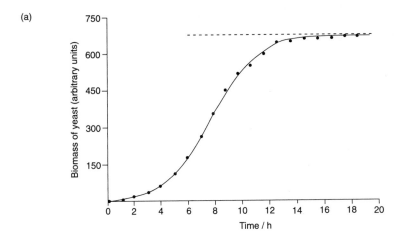

(b)

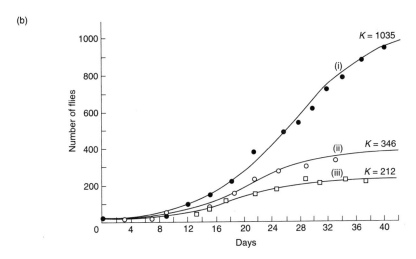

(c)

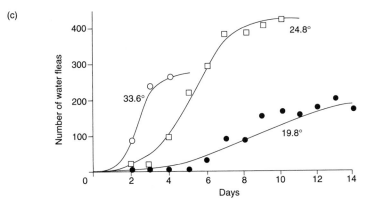

Box 3.2 Simultaneous equations

A linear equation such as:

$5x + 3 = 4$

can be solved for x by simple algebraic manipulation ($x = 1/5$). However, if there are two unknowns in one equation:

$5x + 3 = y$ 　　　　　　　　　　　　　　　　　　　　　　(1)

the equation cannot be solved (for x or y). It can only be solved if there is a second equation in which the same two unknowns (x and y) are represented. For example:

$8 - 4x = y$ 　　　　　　　　　　　　　　　　　　　　　　(2)

Equations 1 and 2 are known as a pair of simultaneous (linear) equations. Note that if there are three unknowns, then three simultaneous equations are required to find the solutions.

There are several methods of determining the values of x and y, as outlined below.

(i) Substitution and elimination

This is achieved by taking one of the equations, expressing one variable (unknown) in terms of the other variable and any coefficients or constants and then substituting for that variable in the second equation. For example, in Eqn 1 we have already that $y = 5x + 3$. We can therefore substitute $5x + 3$ for y in Eqn 2:

$8 - 4x = 5x + 3$

$8 - 3 = 5x + 4x$

$5 = 9x$

$x = 5/9$

Having obtained the value of the first variable (x) this can be substituted into either of the equations to find the value of the second variable (y). For example, substituting $x = 5/9$ into Eqn 1:

$5 \times 5/9 + 3 = y$

$25/9 + 3 = y$

$(25 + 27)/9 = y$

$52/9 = y$

Continued

Box 3.2 (*continued*)

The simultaneous equations have now been solved for x and y. To check these values they could both be substituted into the second equation:

$8 - 4 \cdot 5/9 = 52/9$

$8 - 20/9 = 52/9$

$72/9 - 20/9 = 52/9$ Correct, i.e. LHS = RHS.

(ii) Graphical method
This method only gives an approximate answer but is useful in more complex examples, for example, with a quadratic and linear function. Each linear equation can be represented on a graph as a straight line according to the format $y = mx + c$ where m is the gradient and c is the y-intercept (the point where the line crosses the y-axis). The values of x and y at the intersection give the solution of the equations (Fig. 1). This method was used in the Ricker–Moran plots when the line of unity $N_{t+1} = N_t$ intersected with the discrete logistic function to give the

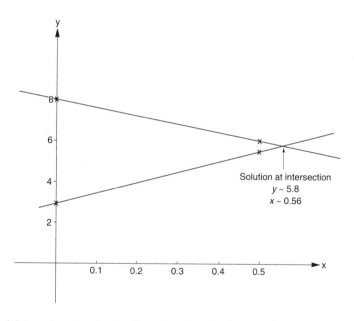

Fig. 1 Intersection of $y = 5x + 3$ and $y = -4x + 8$ yielding the approximate solutions of these two simultaneous equations.

Continued on page 64

Box 3.2 (*continued*)

equilibrium value and for the solution of the quadratic function (intersection of the function with $y = 0$, see Box 2.3).

(iii) Matrix method
See Chapter 4 for details of solution of simultaneous equations using matrix methods.

Question 3.2

Using a simplified (two-parameter) version of Pearl and Reed's equation:

$$N = \frac{b\,e^{at}}{1 + e^{at}}$$

show that the derivative (dN/dt) is the logistic equation.

3.3 Harvesting and maximum sustainable yield

Much of our understanding of the harvesting of wild populations has come from the fisheries literature, with classic long-term studies coming from the North Sea, the Atlantic and the Pacific (see Krebs 1994 for examples). These studies have used both continuous and discrete time population models. These have been complemented by studies on terrestrial populations including high profile and potentially exploitable animals such as the African elephant. In this section we introduce a few basic ideas that lie at the heart of the fisheries and other harvesting studies.

We will assume that population growth is continuous and described by the logistic equation. Although in many cases the logistic equation is too simple a description of population change, it provides a useful entry point to understanding the dynamic possibilities of harvesting. We begin by plotting the rate of population growth, dN/dt, against population size, N_t (Fig. 3.6).

You should note that the curve in Fig. 3.6 is a parabola, i.e. the shape characteristic of a quadratic equation (Box 2.3; the reason for this is seen by expanding the RHS of the logistic equation (Eqn 3.7) to give $rN_t - rN_t^2/K$. You may recall the quadratic form of the discrete logistic, Eqn 2.8). All population sizes which yield values of $dN/dt > 0$ can, in theory, be subject to harvesting. Thus, a fraction of any growing population should be able to be harvested without causing that population to become extinct. The **maximum sustainable yield** occurs where the maximum

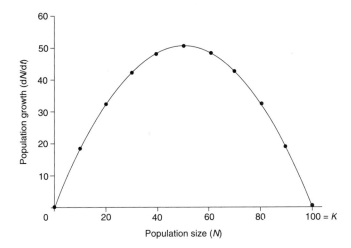

Fig. 3.6 Population growth dN/dt described by the logistic equation against population size (N). K is the carrying capacity (in this example $K = 100$).

value of dN/dt occurs, i.e. when the population is growing most rapidly. This result is generally true for all population growth curves (see the discussion by May & Watts 1992).

To determine the population density corresponding to the maximum sustainable yield we note that the maximum occurs when the gradient of the curve equals 0 (Box 3.1):

$$\frac{d(dN/dt)}{dN} = 0 \tag{3.8}$$

We will now drop the subscript t to make the equations easier to read (N will still mean the population density at time t). For the logistic curve in Fig. 3.6 the derivative of the expression $rN - rN^2/K$ with respect to N is $r - 2rN/K$. Therefore

$$r\left(1 - 2\frac{N}{K}\right) = 0$$

The solution to this equation is that either $r = 0$ (a trivial solution) or $(1 - 2N/K) = 0$. Rearranging the latter yields the solution that $N = K/2$. So the population is growing at its fastest when it is at half its maximum density (at $N = 50$, Fig. 3.6).

To find the maximum value of dN/dt substitute the size at which the maximum occurs ($K/2$) for N in $dN/dt = rN(1 - N/K)$. This gives $dN/dt = rK/4$. (In Fig. 3.5 dN/dt has a maximum of 50, therefore $50 = r100/4$ and $r = 2$.) This method of finding the maximum value is useful if the function for population growth is more complex than the logistic equation.

To develop these ideas let us assume a general form of population growth equation $dN/dt = f(N)$ which describes the rate of change in population size as some function (f) of population size. If this function $f(N)$ has a maximum value(s) of dN/dt over a certain range of values of N then differentiation can be used to find the maximum value(s). You should note that the criterion of $d(dN/dt)/dN = 0$ (Eqn 3.8) is not, in itself, sufficient to identify a maximum value. It might equally identify a minimum value (Box 3.1). A harvesting function can be incorporated into the general population growth equation (Beddington 1979):

$$\frac{dN}{dt} = f(N) - h(N) \tag{3.9}$$

where $h(N)$ gives the reduction in dN/dt at a particular value of N due to harvesting.

The advantage of the generalized form in Eqn 3.9 is that the population change and harvesting functions can be combined on one graph where they are both related to population size (N) (Figs 3.7, 3.8 and 3.10). We will now consider several harvesting possibilities with a logistic growth curve.

If $h(N)$ is constant and therefore not a function of N, then harvesting is represented by a horizontal line on the graph (Fig. 3.7).

We know that sustainable harvesting can only occur when $dN/dt > 0$, i.e. when $f(N) > h(N)$ in Eqn 3.9. The area on the graph where $f(N) > h(N)$ is shaded in Fig. 3.7. This region lies between $N = 20$ and $N = 80$. What happens to the population at different population sizes? For example, consider a population of size 70. At this population size $f(N) > h(N)$ and therefore population size (N) increases ($dN/dt > 0$, Eqn 3.9). In other words, from population size 70 we move to the right on the

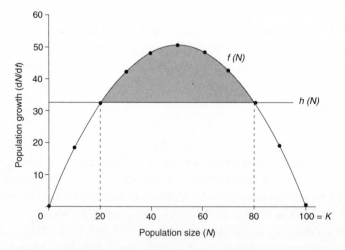

Fig. 3.7 Constant harvesting ($h(N)$) and logistic population growth rate ($f(N)$) plotted against population size (N). K is the carrying capacity. The shaded area shows where $f(N) > h(N)$.

graph. Conversely, for a population of size 90 $f(N) < h(N)$, $dN/dt < 0$ and population size decreases. (From population size 90 we move to the left on the graph.)

By this reasoning it can be seen that the point of intersection of the two lines representing $h(N)$ and $f(N)$ at $N = 80$ (i.e. $h(N) = f(N)$ and therefore $dN/dt = 0$) is a locally stable equilibrium point. A population which receives a small displacement or perturbation away from that equilibrium population size will tend to return to that point. You will probably have noticed a second equilibrium point at $N = 20$ where dN/dt is also equal to 0. $N = 20$ is, however, not a locally stable equilibrium point. If $N < 20$ then $f(N) < h(N)$ and therefore $dN/dt < 0$. Therefore, the population will continue to decrease if reduced below 20. Eventually the population will end up at $N = 0$ and therefore be locally extinct ($N = 0$ is effectively a third equilibrium point which is locally stable). If the population size is increased above $N = 20$ then $f(N) > h(N)$ and therefore the population size will continue to increase until it reaches the stable equilibrium point at $N = 80$.

Thus, even the very simple scenario of constant harvesting combined with logistic growth provides the dynamic possibilities of extinction (below $N = 20$) and local stability (at $N = 80$).

A second possibility for the harvesting function is that it increases linearly with prey population size (Fig. 3.8), for example, the more fish there are, the more people go fishing.

In this example, if $N > 80$ and therefore $f(N) < h(N)$, N will decrease towards 80. If $N > 0$ but less than 80 then dN/dt is positive and N will increase towards 80. Thus $N = 80$ is a stable equilibrium as in the previous example. In both these cases, if the prey population is pushed beyond K (100) then it is predicted from the model

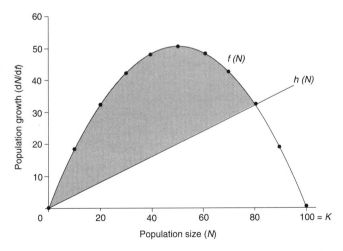

Fig. 3.8 Linear increase in harvesting ($h(N)$) and logistic population growth ($f(N)$), plotted against population size (N).

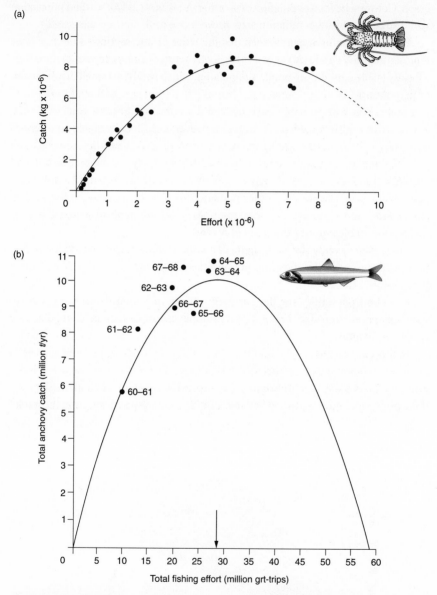

Fig. 3.9 Relationships between fishing effort and catch. Quadratic curves fitted through the data illustrate the problem of estimating maximum sustainable yield. Beddington (1979) discusses the link between catch–effort relationships and the analyses in Figs 3.7 and 3.8. (a) Western rock lobster (Beddington 1979), (b) Peruvian anchovy (Krebs, 1994).

that it will return towards *K*, but it will not stay there as *h(N)* is still greater than *f(N)* and therefore it continues to return to 80.

These analyses of logistic growth combined with different harvesting functions (see also Question 3.3) provide a suite of population stability possibilities, although we do not yet appear to be close to the cyclical dynamics considered in the introduction to this chapter. In the real world logistic (or similar) growth curves can be fitted to fisheries and other data, but the amount of variation around the curve, particularly as the apparent maximum sustainable yield is approached, makes the choice of curve difficult (Beddington 1979; Fig. 3.9). Logistic growth curves have also been used to investigate the harvesting potential of terrestrial populations, such as Neotropical mammals. Any predictions as to the dynamic behaviour of harvested populations needs to be treated with caution, for example, Robinson and Redford recommend using harvesting maxima well below the theoretical maximum sustainable yield.

Question 3.3

Consider a sigmoidal (s-shaped) harvesting function (Fig. 3.10) which, like the logistic growth curve, could occur if harvesting effort is not limited at low population size but is restricted, e.g. by a fixed number of boats, at high densities of fish.

What are the dynamic possibilities arising from this harvesting function coupled with the logistic population growth?

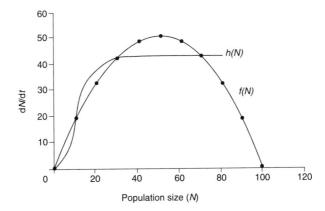

Fig. 3.10 Non-linear increase in harvesting (*h(N)*) and logistic population growth (*f(N)*) plotted against population size (*N*).

3.4 The Lotka–Volterra model of predator and prey dynamics

3.4.1 Construction of the model
In the harvesting model of Section 3.3 the harvesting effort (i.e. predation) was either constant or a simple function of density. This is only likely to be true in a few carefully monitored and controlled examples. It certainly does not represent natural and semi-natural situations where humans are absent or peripheral. However, these types of harvesting model can be thought of as a simplified model of predator–prey interactions in which N is the prey density and $h(N)$ is the reduction in prey population growth due to a predator. For a more informative model of predator–prey interactions we need to have a mathematical expression for the dynamics of the predator. In this section we will describe how models of predator–prey systems in continuous time have been developed and the main results which emerge from them.

The origins of predator–prey models in continuous time are to be found in the independent work of Lotka (1925) in New York and Volterra (1926, 1928) who, in Rome, also derived equations to describe competition (see Chapter 5). The independence of their work is illustrated in their communications to the journal *Nature* from which the opening quotation to this chapter is taken (Lotka 1927). The premise of their predator–prey models was that of 'two associated species, of which one, finding sufficient food in its environment, would multiply indefinitely when left to itself, while the other would perish for lack of nourishment if left alone; but the second feeds upon the first, and so the two species can coexist together' Volterra (1926). Thus, they assumed that prey density (N) increased exponentially (quantified by r_1) in the absence of predators (following Eqn 3.5):

$$\frac{dN}{dt} = r_1 N \tag{3.10a}$$

This was made more realistic by assuming that the change in prey density was described by the logistic equation 3.7, i.e. the prey population would move towards an equilibrium of K in the absence of predation:

$$\frac{dN}{dt} = r_1 N (1 - N/K) \tag{3.10b}$$

(Volterra (1928) considered this possibility, noting that the fluctuations would be damped and the system would tend towards the stationary state (see below). However, he did not pursue this line of enquiry—most of his emphasis in predator–prey systems was on the possibility of cycles using Eqn 3.11 (below) without prey regulation and Eqn 3.12.) Variations on the theme of prey regulation are developed in chapter 5 of May (1981). In the following examples we will use Eqn 3.10b and later briefly consider the results using Eqn 3.10a instead and the problems these results pose.

In the presence of predators the rate of change of prey population size with time, dN/dt, is assumed to be reduced in proportion α to the density of predators (P) multiplied by the density of prey (N):

$$\frac{dN}{dt} = r_1 N(1 - N/K) - \alpha PN$$

or

$$\frac{dN}{dt} = r_1 N - r_1 N_t^2/K - \alpha PN \qquad (3.11)$$

Note that r_1/K may be replaced by a single parameter. As we are now modelling a dynamic system in which the predator population density may also fluctuate, we need to develop an equation for dP/dt. Lotka and Volterra assumed that, in the absence of prey, the predator population size would decline exponentially, quantified by r_2, i.e. they assumed that the predator species specialized on one species of prey:

$$\frac{dP}{dt} = -r_2 P$$

In the presence of prey, this decline would be counteracted by an increase in predator density, again in proportion β to the density of predators (P) multiplied by the density of prey (N):

$$\frac{dP}{dt} = -r_2 P + \beta PN \qquad (3.12)$$

Equations 3.11 and 3.12 provide a system of two **coupled first-order non-linear differential equations** (Box 3.1):

$$\frac{dN}{dt} = r_1 N - r_1 N_t^2/K - \alpha PN$$

$$\frac{dP}{dt} = -r_2 P + \beta PN$$

In Section 3.4.2 we will consider a graphical technique for analysing the behaviour of coupled differential equations. This technique is very useful because systems of differential equations may arise in all types of ecological interaction (including the various manifestations of predator–prey interaction, i.e. plant–herbivore, host–parasitoid and host–pathogen; competitive and mutualistic interactions, see

Chapters 5 and 7). Before employing this technique it is worth considering how we might expect Eqns 3.11 and 3.12 to behave. Let us begin when prey (N) and predator (P) are at low densities. If N and P are both small, then the product NP will be very small. We could therefore ignore NP (assume it has a value of 0), leaving dN/dt with a positive value and dP/dt with a negative value.

Therefore the prey population is increasing and the predator population is decreasing. As the prey population increases (N gets larger) the predator population has something to consume and we expect it to begin growing (i.e. dP/dt to become positive and therefore P to increase). We expect the prey to increase more rapidly than the predator. As the predator population builds up, following the prey population, so the value of the product NP increases. Eventually there comes a point at which αPN is so high that dN/dt becomes negative and the prey density begins to decline. By continuing this line of argument we might predict that the prey and predator densities oscillate with the predator density lagging behind the prey density, i.e. we reach the possibility of cycles of abundance observed in the field. As we shall see this will only occur under particular conditions of the Lotka–Volterra model, and the following mathematical analysis will help develop the rather clumsy verbal reasoning.

Question 3.4

Give an ecological interpretation of the coefficients α and β in Eqns 3.11 and 3.12. Predation causes a loss to the prey population of αPN and a gain to the predator population of βPN. Why is it unlikely that $\alpha = \beta$?

3.4.2 Phase plane analysis

To understand the dynamics produced by the Lotka–Volterra equations (Eqns 3.11 and 3.12) we examine their behaviour on a **phase plane** where the densities of predator and prey at time t are plotted on a graph and linked to their rate of change at those densities. This is a graphical method of formalizing the intuitive argument used above, developed by Rosenzweig and MacArthur (1963) for predator–prey models. A phase plane can be used to describe change in any two variables in coupled differential equations. Descriptions of change in phase space are used for more than two variables.

We will begin at time 1 with a value of P_1 for predators and N_1 for prey (Fig. 3.11). To these densities we attach a vector showing the change in P and N from that point. The vector is a combination of two orthogonal vectors (vectors at right angles)—one representing the value of dP/dt and one representing the value of dN/dt. These two vectors are added to give the resultant vector (Fig. 3.11).

It is not necessary to know the precise direction of change from any given combination of N and P. Indeed, the beauty of the phase plane analysis is that the

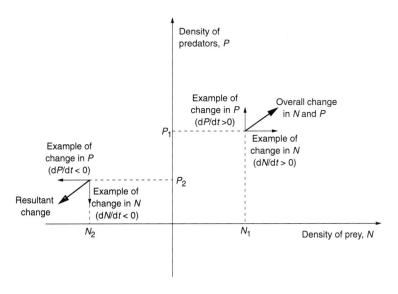

Fig. 3.11 Prey (N) and predator (P) densities and representation of associated dynamics in the phase plane.

dynamics of the system can be understood by sketching some of the vectors based on the signs, i.e. positive or negative values of dP/dt and dN/dt. We will try this with the Lotka–Volterra equations 3.11 and 3.12. Once the values of r, α, β and K have been decided (or determined from the laboratory or field) the values of dN/dt and dP/dt will depend only on the values of prey density N and predator density P—exactly the requirement for the phase plane analysis. We will assume the following parameter values: $r_1 = 3$ and $r_2 = 2$ (prey > predator), $\alpha = 0.1$, $\beta = 0.3$ and $K = 10$. Therefore $dN/dt = 3N - 3N^2/10 - 0.1PN$ and $dP/dt = -2P + 0.3PN$.

Initially, it is helpful to know when $dN/dt = 0$ and when $dP/dt = 0$ (known as zero growth isoclines) i.e. there is no change in N or P respectively. First, $dN/dt = 0$ when $3N - 3N^2/10 - 0.1PN = 0$ or $(3 - 3N/10 - 0.1P)N = 0$. Thus, either $N = 0$ (the trivial solution in which prey are absent) or $3 - 3N/10 - 0.1P = 0$. The latter can be rearranged to give:

$$P = 30 - 3N$$

This is a straight-line equation which can be plotted on the phase plane (Fig. 3.12).

Similarly $dP/dt = 0$ when $-2P + 0.3PN = 0$ or $P(-2 + 0.3N) = 0$. The solutions are either $P = 0$ (the trivial solution) or $-2 + 0.3N = 0$, which can be rearranged to give $N = 6.667$.

This solution is plotted on the phase plane (Fig. 3.12) as a vertical line at $N = 6.7$. The intersection of the two lines representing $dN/dt = 0$ and $dP/dt = 0$ is important, as this is where neither P nor N changes in density—this is an equilibrium point

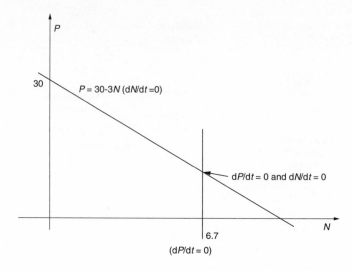

Fig. 3.12 Lines representing no change in prey density dN/dt = 0 and predator density dP/dt = 0 on a phase plane.

whose stability will be seen to be of great significance to the dynamics of the predator–prey system. The phase plane has been divided up into four regions, produced by the intersection of dN/dt = 0 and dP/dt = 0. Each region is characterized by a particular combination of positive or negative values of dN/dt and dP/dt (Fig. 3.13).

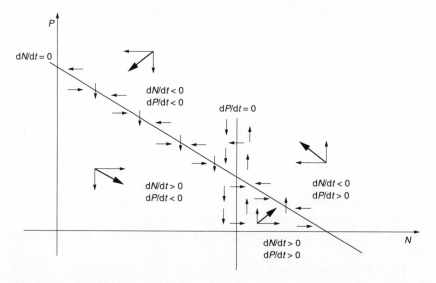

Fig. 3.13 Overall direction of change in prey and predator densities in four regions of the phase plane.

This is helpful as any point (N, P) in a given region will have a particular combination of vectors attached to it. Although the relative magnitudes of the vectors will depend on where the points are in the region, the overall direction of change in N and P (represented by the resultant vector) will always be the same. These directions are indicated in bold in Fig. 3.13.

To find the directions of the vectors in any region we need to determine whether dN/dt and dP/dt are positive or negative in that region. For dP/dt, if $N > 6.7$ then $-2 + 0.3N > 0$ and dP/dt is positive. If $N < 6.7$ then dP/dt is negative. Therefore to the left of the vertical line $dP/dt = 0$ (at $N = 6.7$) all the change in P is negative (arrows point down on Fig. 3.13). To the right of the vertical line $dP/dt > 0$ and the arrows (vectors) point up. When $dP/dt = 0$ there is no change in P, so there can only be change in N, as indicated by the horizontal arrows. (The direction of the horizontal arrows is only known when the regions of dN/dt greater or less than zero have been determined.)

Now consider the two regions either side of $dN/dt = 0$. Where is $dN/dt < 0$? $3 - 3N/10 - 0.1P < 0$ giving $30 - 3N < P$. This occurs above the line of $dN/dt = 0$. You can check this by substituting values for N and P, e.g. $P = 40$ and $N = 0$: $30 - 0 < 40$ is true.

Therefore above the line of $dN/dt = 0$ the horizontal arrows point to the left, i.e. $dN/dt < 0$ whilst below the line they point to the right. When $dN/dt = 0$ there is no change in N so there can only be change in P (indicated by vertical arrows, with the direction determined by whether they are to the left or right of $dP/dt = 0$).

Now, for any point (N, P) in the phase plane we know the direction of change. Taking any starting point on the phase plane it is possible to look at the dynamics of N and P as a trajectory across the phase plane. With the above parameter values, starting at this point sends the population spiralling rapidly into the equilibrium point (Fig. 3.14), i.e. in this example we get a stable equilibrium point rather than sustained cycles. You should try this as well, starting from a different point. The key to drawing the trajectory is to know the general direction you need to go, e.g. between 3 and 6 o'clock in the left-hand, bottom corner of the phase plane, and the horizontal or vertical direction to be in when you cross the next equilibrium line. Some example instructions are given in Fig. 3.14.

If the $dP/dt = 0$ line is kept vertical and the slope of $dN/dt = 0$ is altered, what is the effect on dynamics and stability? In particular what happens when $dN/dt = 0$ (horizontal) which is equivalent to removing the prey density dependence, i.e. by using the unrealistic equation 3.10a minus αPN. With $dN/dt = 0$ starting at any point simply sends the trajectory on an elliptical path back to where it started (Fig. 3.15a). This is the case of **neutral stability** for which the Lotka–Volterra model received much criticism. Neutral stability, akin to a frictionless pendulum, is referred to as a structurally unstable model (May 1973a), in which the amplitude of cycles is determined by initial conditions and the cycles persist with unchanging amplitude.

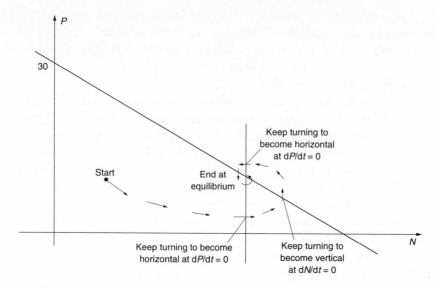

Fig. 3.14 An example of a plot of the trajectory of predator and prey dynamics on a phase plane.

This is in contrast to the limit cycles discussed in Chapter 2 where the cycles fluctuate between particular densities regardless of the initial conditions, i.e. the starting density. If the gradient of $dN/dt = 0$ is positive then the trajectory spirals out ending in local extinction of predator and prey (Fig. 3.15b). If the gradient of $dN/dt = 0$ is negative then the trajectory spirals in to a stable equilibrium (Fig. 3.15c and as in Fig. 3.14). Further examples are given in Maynard Smith (1974) and Crawley (1992).

In fact, with the Lotka–Volterra equations, the only cycles produced are neutrally stable. For structurally stable cycles we need a time delay of some description, which may be expressed either as a delay differential equation (Hutchinson 1948) or, as in the next section (3.5), by coupled difference equations.

3.5 Delayed density dependence

3.5.1 Second-order non-linear difference equations

We will now consider a discrete version of the Lotka–Volterra model (May 1973b). By analogy with Eqns 3.11 and 3.12:

$$N_{t+1} = \lambda_N N_t (1 - N_t/K) - a P_t N_t \tag{3.13}$$

$$P_{t+1} = \lambda_P P_t + b P_t N_t \tag{3.14}$$

Therefore, in the absence of predation the prey (Eqn 3.13) are self-regulated according to the discrete logistic equation (Eqn 2.8), and in the absence of prey the

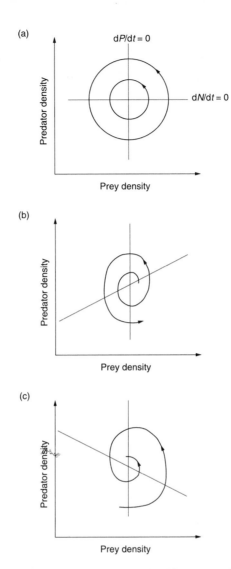

Fig. 3.15 Effect of changing gradient of $dN/dt = 0$. Reprinted from Maynard Smith (1974). (a) Neutrally stable cycles of predators and prey (gradient of $dN/dt = 0$ is zero). (b) Densities of predator and prey spiral out towards extinction (gradient of $dN/dt = 0$ is positive). (c) Densities of predators and prey spiral into a stable equilibrium (gradient of $dN/dt = 0$ is negative).

predators decline at the rate of λ_P (which will be less than 1). If the time subscript for the predator equation (3.14) is reduced on both sides by $t = 1$ to give:

$$P_t = \lambda_P P_{t-1} + bP_{t-1} N_{t-1}$$

and the RHS substituted for P_t in Eqn 3.13:

$$N_{t+1} = \lambda_N N_t(1 - N_t/K) - a\,(\lambda_P P_{t-1} + bP_{t-1}N_{t-1})N_t \tag{3.15}$$

then we are left with a **second-order non-linear difference equation** (Box 2.1), because N_{t+1} is explained by terms which are two time steps (P_{t-1} and N_{t-1}) earlier (in addition to a term one time step earlier, N_t).

Therefore, a pair of coupled first-order difference equations is equivalent to a second-order difference equation. The dynamics produced by second-order non-linear difference equations are very interesting and quite different from their first-order cousins. Second-order equations (or their coupled first-order equivalents) can produce cycles which are similar to those of predators and prey observed in the field. As an example consider the dynamics of the larch bud-moth (*Zeiraphera diniana*), which periodically defoliates its host tree larch (*Larix decidua*), and the pine looper moth (*Bupalus piniaria*), which specializes on Scots pine (*Pinus sylvestris*). Detailed sampling of the larvae of the larch bud-moth has shown that the optimum area for survival and fecundity is the subalpine region of the Swiss alps between 1700 and 2000 m (Baltensweiler 1993). In this region the larch bud-moth reaches carrying capacity within 4–5 generations and has cycle lengths of 8–9 years (Fig. 3.1). The clear cycles of *Zeiraphera* are in contrast to the rather irregular and sometimes absent cycles of *Bupalus* (Fig. 3.16).

Defoliation of the larch in one year affects the tree's physiology in the subsequent year, including a reduction in the quality of the food for the larvae which, in turn,

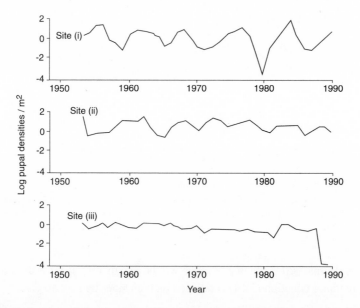

Fig. 3.16 Cycles of population abundance of Pine looper (*Bupalus piniaria*) at three sites (Broek huizen *et al.* 1993).

triggers the collapse of the insect populations in the following years. Alteration in food quality is seen in shorter needles, decreased nitrogen content and greater fibre content. Thus one cause of the population cycles is believed to be build-up of larval densities beyond carrying capacity, leading to defoliation and therefore reduced food in the next year. (Interaction with the granulosis virus is also believed to produce population cycles in the larch bud-moth—see Chapter 7 for models of host–virus interaction.) This is an example of **delayed density dependence** in which the effects of density dependence are carried over to the next generation. Delayed density dependence, first described by Varley (1947), underpins the dynamics of many predator–prey systems and can be described by second-order (and higher-order) non-linear difference equations. Broekhuizen *et al.* (1993) identified a related delayed density-dependent mechanism by which population cycles may be produced in the pine looper moth:

> There is a strong negative correlation between annual growth increment of *Pinus sylvestris* in Tentsmuir forest and the pupal density of *Bupalus piniaria* in the previous two years (Straw 1991). This indicates that *B. piniaria* may have a substantial influence upon their host trees' physiologies. This may 'feed back' upon the *B. piniaria* population such that the one year's *B. piniaria* influence the reproductive success of the next year's population.

The *B. piniaria–P. sylvestris* interaction therefore provides a second example of delayed density dependence which is believed to be central to the large fluctuations and possible cycles of abundance of predators (in this case insect herbivores) and their prey (in this case forest trees). Note that in this and the *Zeiraphera* examples the prey change in the abundance of biomass or the amount of components of biomass, such as nitrogen content, rather than in numbers or density.

To explore delayed density dependence further we will look at a simple version of a second-order non-linear difference equation which can be parameterized from census data. This covers the work of Turchin (1990, see also Turchin & Taylor 1992) who used a second-order Ricker equation to investigate the likelihood of delayed density dependence amongst herbivorous forest insects including *Zeiraphera* and *Bupalus*. We met the first-order version of the Ricker equation in Chapter 2 (Eqn 2.13) and will use the same regression technique to estimate parameters. The second-order version is:

$$N_{t+1} = N_t \, \mathrm{e}^{(r + aN_t + bN_{t-1})} \tag{3.16}$$

The parameter a represents the strength of direct density dependence whilst b represents the strength of delayed density dependence. Note that this is equivalent to $N_{t+1} = N_t \, \mathrm{e}^r \, \mathrm{e}^{(aN_t + bN_{t-1})} = \lambda N_t \, \mathrm{e}^{(aN_t + bN_{t-1})}$. If $b = 0$ then Eqn 3.16 reverts to the first-order equation. For regression purposes we also assume an unexplained variance term in the model (see Chapter 2). To estimate the parameters a and b we rearrange Eqn 3.16 (divide by N_t and take natural logs) to give:

$$\ln \frac{N_t + 1}{N_t} = r + aN_t + bN_{t-1} \tag{3.17}$$

$\ln (N_{t+1}/N_t)$ can then be regressed against N_t and N_{t-1}. This was the method used by Turchin (1990) who showed significant delayed density dependence (i.e. values of b significantly different from 0) in 10 out of 14 data sets. In considering the significance of the parameters we should recall from Chapter 2 the possibility of overestimating the significance of a or b using this regression method. The values of a, b and r for *Bupalus* and *Zeiraphera* are given in Table 3.1.

The dynamics resulting from the estimated parameter values can be found by simulation using Eqn 3.16 (Fig. 3.17). You will see that the predicted dynamics for the two species are quite different with *Bupalus* predicted to be stable with an equilibrium population size of about 150. *Zeiraphera*, in contrast, is predicted to cycle with periods of 6–7 years, slightly shorter than the 8–9 year cycles observed in the field. These results from the deterministic equation 3.16 are in agreement with the clearly defined cycles of *Zeiraphera* (Fig. 3.1b) and the poorly defined cycles of *Bupalus* (Fig. 3.16). In other words, analysis of Eqn 3.16 has suggested a much stronger deterministic signal for cycle production in *Zeiraphera* compared with *Bupalus*.

Simulation of another second-order model in Broekhuizen *et al.* also suggests a stable equilibrium for other *Bupalus* populations. The critical parameter determining cyclical behaviour in Eqn 3.16 is r (you could explore this with simulations). Indeed, whilst 10 out of 14 of Turchin's data sets showed delayed density dependence, only three have values of r high enough to produce cycles. Turchin & Taylor (1992) noted that focusing on direct density dependence rather than delayed density dependence leads to potentially misleading results. They gave the example of the analysis of Hassell *et al.* (1976) who estimated direct density dependence and then looked at the stability (or otherwise) of insect population dynamics (Box 2.4).

Table 3.1 Parameter values for r, direct (a) and delayed (b) density dependence derived from linear regression of census data of *Bupalus* and *Zeiraphera* using Eqn 3.17 (Turchin 1990). Both values of a and b are significantly different from 0. The final column gives the percentage of variance explained by the regression model.

	r	a	b	%var
Bupalus	0.34	−0.0005	−0.0018	24.1
Zeiraphera	1.20	−0.0001	−0.020	1.1

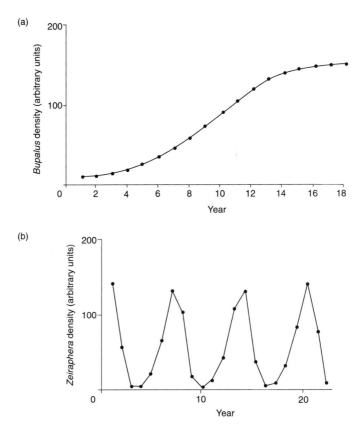

Fig. 3.17 Predicted dynamics of (a) *Bupalus* and (b) *Zeiraphera* using Eqn 3.16 and the parameter values in Table 3.1. The figure for *Bupalus* shows the population moving smoothly (and sigmoidally) from initial conditions of $N_1 = 10$ and $N_2 = 10$ towards a stable equilibrium. The figure for *Zeiraphera* is shown once the population has settled down from its initial conditions.

Hassell *et al.* classified the larch bud-moth in the Engadine Valley, Switzerland as stable, although this population was according to Turchin and Taylor 'arguably the most convincing example of a … cyclical system in our data set'.

Question 3.5

Calculate the value of the equilibrium population (N_e) for *Bupalus* from Eqn 3.16 and check it against the approximate value in Fig. 3.17.

3.6 Conclusions

Why do we want to model the dynamics of predators and their prey? There are two related answers to this question. First, the regulation of numbers of either predators or prey will depend critically upon the nature of the interaction and the extent of self-regulation of predators and prey. More than 60 years on from Lotka and Volterra ecologists are still not certain whether predation or competition (or both or neither) regulate populations (Crawley 1992). Models allow us to probe into the possibilities for regulation and to explore their implications for dynamics of predators and prey. We have seen examples in this chapter where tweaking of extremely simple models can result in stable oscillations or equilibrium, population decline or increase. Second, these theoretical problems have important applications. Harvesting of plants and animals and introduction of biological control agents are two subject areas where direct and possibly large-scale economic and environmental benefits can be accrued from investigation of models of predator–prey dynamics.

CHAPTER 4

Population dynamics of species with complex life-histories

We now have an understanding of the dynamics of populations described by first-order difference and differential equations (Chapters 2 and 3, respectively) and have begun to explore the interactions between such populations, introducing second-order equations (Chapter 3). We are now ready to consider more complex examples developed from these models. Here are a number of possible developments.

1 Incorporation of **age, stage or size structure** into the models. The first-order difference equations in Chapter 2 described the dynamics of 'annual' organisms in which all individuals were the same age. Similarly the continuous time equations in Chapter 3 took no account of differences between individuals. In the present chapter we will consider the development of models of populations composed of individuals of different age, stage or size.

2 Investigation of community dynamics and structure produced by interactions between populations. These include within trophic level (competition, mutualism) and between trophic level interactions (plant–herbivore, predator–prey, host–parasite). Predator–prey interactions were discussed in Chapter 3 and will be developed with other interactions in Chapter 5. Host–virus interactions are the focus of Chapter 7, embracing many of the modelling techniques of the previous chapters.

3 Construction of spatially structured models in which local populations, whose dynamics may be described by first-order equations, are linked by dispersal. The effects of spatial structure are considered in Chapter 6.

4.1 Analysis of age-structured populations

4.1.1 Introduction

In this chapter we will use **matrices** to summarize the structure and parameters of a population composed of organisms with complex life-histories. Basic matrix operations are described in Box 4.1. The use of matrices is covered in rich detail in Caswell (1989), which should be read not only for its treatment of matrix models but also for its insights into the relationship between models and applications. Individuals of long-lived species may have widely varying patterns of pre-reproductive and reproductive life (Fig. 4.1). We will assume that, although generations overlap, reproduction occurs at particular times of year and therefore discrete time models are appropriate. From the point of view of population dynamics there are two

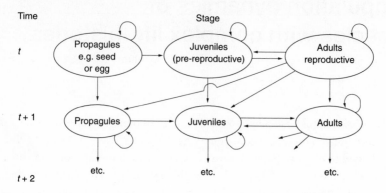

Fig. 4.1 Representation of life-histories of plant and animal species. Arrows show all possible transitions between stages, within and between years. Production of juveniles from adults includes vegetative reproduction in plants.

important differences between long-lived organisms with overlapping generations and annual or short-lived organisms with separate generations. First, long-lived organisms may delay reproduction for one or more years and, second, they may survive after reproduction to reproduce again. Thus in this category we consider monocarpic or semelparous (single-flowering) and polycarpic or iteraparous (multiple-flowering) perennial plants, many vertebrate and some invertebrate species. In all of these cases the life-history of an individual may be divided up in terms of its age (e.g. time of first reproduction), stage (e.g. adult or juvenile) or size (e.g. only plants over a certain size will reproduce).

Box 4.1 Matrices

Construction of a matrix
A matrix is an ordered rectangular array of numbers, for example, the matrix:

$$\begin{pmatrix} 2 & 3 \\ 5 & 7 \end{pmatrix}$$

contains four elements (2, 3, 5, 7) in two rows and two columns. A matrix may have any number of rows or columns. A matrix is described by its number of rows and columns, for example, the matrix:

$$\begin{pmatrix} 1 & -2 & 7 \\ 3 & 5 & 1 \end{pmatrix}$$

has two rows and three columns and is therefore a 2×3 matrix. A matrix with

Continued

Box 4.1 (*continued*)

one column is referred to as a column matrix or column vector. A matrix with equal numbers of rows and columns is called a square matrix. Matrices can be added, subtracted, multiplied and multiplied by a scalar (a single number, i.e. a 1×1 matrix!).

Matrix addition and subtraction

Matrices can only be added or subtracted if they have the same number of rows and columns because it is the elements which are added or subtracted, for example:

$$\begin{pmatrix} 2 & 3 \\ 5 & 7 \end{pmatrix} + \begin{pmatrix} 1 & 4 \\ 5 & 8 \end{pmatrix} = \begin{pmatrix} 3 & 7 \\ 10 & 15 \end{pmatrix}$$

Subtraction operates in exactly the same way, element by element.

Matrix multiplication

An example will illustrate the process:

$$\begin{pmatrix} 2 & 3 \\ 5 & 7 \end{pmatrix} \begin{pmatrix} 1 & 4 \\ 5 & 8 \end{pmatrix} = \begin{pmatrix} 2 \times 1 + 3 \times 5 & 2 \times 4 + 3 \times 8 \\ 5 \times 1 + 7 \times 5 & 5 \times 4 + 7 \times 8 \end{pmatrix}$$

$$= \begin{pmatrix} 17 & 32 \\ 40 & 76 \end{pmatrix}$$

Thus, to give 17 the elements in the first row of the first matrix are multiplied by the corresponding elements in the first column of the second matrix. The fact that top row and first column have been multiplied together gives the position of top row, first column for the product. Unlike matrix addition or subtraction it is possible to multiply unequal matrices. For example:

$$\begin{pmatrix} 2 & 3 \\ 5 & 7 \end{pmatrix} \begin{pmatrix} 1 \\ 5 \end{pmatrix} = \begin{pmatrix} 2 \times 1 + 3 \times 5 \\ 5 \times 1 + 7 \times 5 \end{pmatrix} = \begin{pmatrix} 17 \\ 40 \end{pmatrix}$$

In most ecological applications we are concerned with the multiplication of square matrices or the multiplication of a square matrix by a column matrix.

Multiplication by scalar (single value)

In this case all elements are multiplied by a single value, for example:

$$3\begin{pmatrix} 2 & 3 \\ 5 & 7 \end{pmatrix} = \begin{pmatrix} 6 & 9 \\ 15 & 21 \end{pmatrix}$$

Continued on page 86

Box 4.1 (*continued*)

Identity matrix, inverse and determinant

The identity matrix depends on the operation (addition or multiplication) under consideration. The identity matrix is the one which leaves the other matrix unchanged under a particular operation. For multiplication of a 2×2 matrix the identity matrix is:

$$\begin{pmatrix} 1 & 0 \\ 0 & 1 \end{pmatrix}$$

The inverse matrix is a matrix which when multiplied by another matrix yields the identity. The inverse of a 2×2 matrix is found as follows. Consider the matrix:

$$\begin{pmatrix} 2 & 3 \\ 4 & 7 \end{pmatrix}$$

Begin by calculating the determinant, which is the elements 3 and 4 multiplied together and subtracted from the product of the elements 2 and 7.

$$\begin{vmatrix} 2 & 3 \\ 4 & 7 \end{vmatrix} = 2 \cdot 7 - 3 \cdot 4 = 2$$

Note that the determinant of a matrix is indicated by parallel lines. Now swop the elements in the leading diagonal (2 and 7) and change the sign of the other two:

$$\begin{pmatrix} 7 & -3 \\ -4 & 2 \end{pmatrix}$$

Finally multiply each element by the reciprocal of the determinant to give the inverse matrix:

$$\frac{1}{2}\begin{pmatrix} 7 & -3 \\ -4 & 2 \end{pmatrix} = \begin{pmatrix} \frac{7}{2} & \frac{-3}{2} \\ \frac{-4}{2} & \frac{2}{2} \end{pmatrix}$$

To check that the inverse matrix is correct, multiply the inverse by the original matrix. This should give the identity matrix. For the above example:

Continued

Box 4.1 (*continued*)

$$\begin{pmatrix} \frac{7}{2} & \frac{-3}{2} \\ \frac{-4}{2} & \frac{2}{2} \end{pmatrix}\begin{pmatrix} 2 & 3 \\ 4 & 7 \end{pmatrix} = \begin{pmatrix} 7-6 & 10.5-10.5 \\ -4+4 & -6+7 \end{pmatrix} = \begin{pmatrix} 1 & 0 \\ 0 & 1 \end{pmatrix}$$

4.1.2 An age-structured population

Imagine a species, the individuals of which breed once a year, starting at age 3 years and which live to a maximum of 5 years. The reproduction and survival of these organisms can be described by a set of first-order difference equations. These give either the survival of individuals of different age or the reproductive output of individuals aged 3–5. Assume initially that the age-specific fecundity and survival parameter values are density-independent and are constant from year to year. For example, survival from birth to age 1 is described as:

Number of   Number
individuals aged 1 = born (age 0) × $\dfrac{\text{Fraction surviving}}{\text{from age 0 to 1}}$
(in year $t + 1$)   (in year t)

This can be represented in mathematical notation:

$$N_{1,t+1} = N_{0,t}s_{0,1} \tag{4.1}$$

In Eqn 4.1 the double subscript for the number of individuals (N) indicates the age-class and the time (year). For the survival parameter (s) the double subscript describes the ages over which survival is considered. We can write similar equations describing the survival for the other age-classes:

$$N_{2,t+1} = N_{1,t}s_{1,2} \tag{4.2}$$

$$N_{3,t+1} = N_{2,t}s_{2,3} \tag{4.3}$$

$$N_{4,t+1} = N_{3,t}s_{3,4} \tag{4.4}$$

$$N_{5,t+1} = N_{4,t}s_{4,5} \tag{4.5}$$

The fraction of individuals surviving from birth (age 0) to age 5 is therefore the multiple of the separate survival values from ages 0 to 1, 1 to 2 and so on, i.e. $s_{0,1}s_{1,2}s_{2,3}s_{3,4}s_{4,5}$. We will assume that any individuals surviving to reproduce aged 5 then die. Therefore, for any given value of N_0, N_5 could be predicted.

An equation is also required for the production of offspring (age 0 individuals in year t) by individuals aged 3–5 in the same year (t):

$$N_{0,t} = N_{3,t}f_3 + N_{4,t}f_4 + N_{5,t}f_5$$

f_3, f_4 and f_5 are age-specific fecundity parameters representing the average number of offspring per individual of that age in year t. The RHS of this equation can be multiplied by $s_{0,1}$ (Eqn 4.1) to give the predicted number of offspring surviving to age 1 in year $t + 1$:

$$N_{1,t+1} = N_{3,t}s_{0,1}f_3 + N_{4,t}s_{0,1}f_4 + N_{5,t}s_{0,1}f_5 \tag{4.6}$$

Equations 4.2–4.6 provide a complete description of the density-independent survival and fecundity of individuals in this age-structured population. In order to explore by simulation the dynamics of this age-structured population we could use Eqns 4.2–4.6, incorporate them into a computer program and run the model, given certain initial values. Alternatively we can employ analytical techniques, in which case it is helpful to rewrite the equations in a different form, employing a matrix structure.

Question 4.1

There are no density-dependent processes in this model so, by analogy with Eqn 2.1 in Chapter 2 and Eqn 3.5 in Chapter 3, how do you expect the model population to behave over time?

4.1.3 Matrix description of an age-structured population

Matrices allow us to simplify the representation of a set of first-order equations such as Eqns 4.2–4.6:

$$
\begin{pmatrix} N_1 \\ N_2 \\ N_3 \\ N_4 \\ N_5 \end{pmatrix}
=
\begin{pmatrix}
0 & 0 & s_{0,1}f_3 & s_{0,1}f_4 & s_{0,1}f_5 \\
s_{1,2} & 0 & 0 & 0 & 0 \\
0 & s_{2,3} & 0 & 0 & 0 \\
0 & 0 & s_{3,4} & 0 & 0 \\
0 & 0 & 0 & s_{4,5} & 0
\end{pmatrix}
\begin{pmatrix} N_1 \\ N_2 \\ N_3 \\ N_4 \\ N_5 \end{pmatrix}
\tag{4.7}
$$

$$\mathbf{v}_{t+1} \qquad\qquad \mathbf{M} \qquad\qquad \mathbf{v}_t$$

Three matrices (Eqn 4.7) are required to summarize the five difference equations. First, two column vectors (matrices) representing the numbers of individuals at ages 1–5 at $t + 1$ and t ($\mathbf{v}_{t+1}$ and $\mathbf{v}_t$, respectively). The column vector is often referred to as the **population structure vector** or **age distribution vector**. Note that matrix symbols are given in bold. Second, there is a square matrix, **M** which gives all the fecundity and survival values. This is often referred to as the **population projection matrix**. You may wish to multiply out the matrix **M** and the population vector $\mathbf{v}_t$ to check that they agree with Eqns 4.2–4.6 (Box 4.1 gives details of matrix multiplication). For example, N_1 in $\mathbf{v}_{t+1}$ is equal to the first row of **M** $(0, 0, s_{0,1}f_3, s_{0,1}f_4, s_{0,1}f_5)$ multiplied by the corresponding elements of the column matrix $(N_1, N_2, N_3, N_4, N_5)$ to give Eqn 4.6. Representation of age-structured populations in this manner was first described by Bernardelli (1941), Lewis (1942) and Leslie (1945, 1948).

Matrix equations such as Eqn 4.7, representing difference equations, can be written in a general form to describe any age-structured population:

$$\mathbf{v}_{t+1} = \mathbf{M}\mathbf{v}_t \qquad (4.8)$$

where $\mathbf{v}_t$ and $\mathbf{v}_{t+1}$ are population vectors of the numbers of individuals at different ages (or sizes or stages) at t and $t + 1$, respectively, and **M** is a square matrix in which the number of columns and rows is equal to the number of age-classes. (You will see the similarity of Eqn 4.8 to Eqn 2.1, $N_{t+1} = \lambda N_t$. This similarity is considered below.)

4.1.4 Determination of the eigenvalue and eigenvector
In order to proceed with the analytical investigation we will take a much simpler age-structured population and then discuss more complicated examples in the light of results from the simpler version.

Let us consider a population of biennial plants (Fig. 4.2). The plant population has two age-classes which correspond to particular developmental stages. In the first year the plant forms rosettes from overwintering seed. In the second year

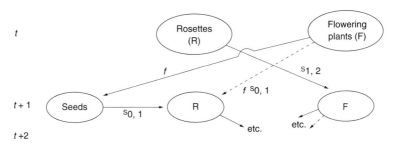

Fig. 4.2 Representation of biennial with f, $s_{0,1}$ and $s_{1,2}$ shown, following Fig. 4.1 and Eqns 4.9 and 4.10. Seed is not included in the model and the transition from F_t to R_{t+1} effectively occurs in the dashed arrow.

the surviving rosettes flower. Therefore, this model could also be described as a stage-structured population (Lefkovitch 1965) composed of rosettes and flowering plants. Manly (1990, ch.6) provides a review of matrix models of stage-structured populations. It is a coincidence in this case that each stage survives for one unit of time—in most cases this would not be true, e.g. a tree species may spend many years at one defined stage.

We will assume that the plant is a strict biennial, i.e. it always flowers in the second year (assuming it survives) and always dies after flowering. Relaxing these assumptions would allow us to model populations of species where flowering is delayed beyond the second year and where vegetative reproduction may occur after flowering. We can summarize the dynamics of the population with two first-order equations:

$$R_{t+1} = F_t f s_{0,1} \tag{4.9}$$

$$F_{t+1} = R_t s_{1,2} \tag{4.10}$$

where R is the number or density of rosette plants, F is the number of flowering plants, f is the average number of viable seed per flowering plant, $s_{0,1}$ represents the fraction of seed surviving between dispersal from the mother plant to rosette formation and $s_{1,2}$ describes the fraction of rosettes surviving until flowering (after which all remaining plants are assumed to die).

In constructing such models it is often the case that stages such as seed are omitted—this will depend on the units of time chosen and the census time. For example, we could have looked at changes from spring to autumn and autumn to spring in which case seed may need to have been included, or at least a seed/small rosette stage. In constructing difference equation models we should be aware of this possibility and consider the implications for the predicted dynamics.

As before, it is possible to write Eqns 4.9 and 4.10 in matrix notation (the algebraic shorthand for the matrices is indicated below them):

$$\begin{pmatrix} R \\ F \end{pmatrix} = \begin{pmatrix} 0 & fs_{0,1} \\ s_{1,2} & 0 \end{pmatrix} \begin{pmatrix} R \\ F \end{pmatrix} \tag{4.11}$$

$$\mathbf{v}_{t+1} \qquad\qquad \mathbf{M} \qquad\quad \mathbf{v}_t$$

We will now describe a mathematical analysis which will reveal two important results. First, it will provide the ratio of R to F, i.e. the composition or structure of the population. Second, it will give us the finite rate of change of the biennial population. This is equivalent to the finite rate of change (λ) described in Chapter 2. Therefore this analysis makes an important assumption about the square matrix, $\mathbf{M}$. It assumes that the matrix can be replaced by a single value (λ), in other words $\mathbf{M}\mathbf{v}_t = \lambda\mathbf{v}_t$. If this is true, then the matrix equation 4.11 (and the more general equation 4.8) can be written as the density-independent equation 2.1 in Chapter 2, except now $\mathbf{v}_{t+1}$ and $\mathbf{v}_t$ are population vectors rather than single numbers:

$$\mathbf{v}_{t+1} = \lambda \mathbf{v}_t \tag{4.12}$$

You should note that in multiplying the vector, $\mathbf{v}_t$, by λ, that all elements of the matrix are multiplied by λ. (λ is a scalar, Box 4.1.) Equating the RHS of Eqns 4.11 and 4.12, i.e. values at time t, we have:

$$\begin{pmatrix} 0 & fs_{0,1} \\ s_{1,2} & 0 \end{pmatrix} \begin{pmatrix} R \\ F \end{pmatrix} = \lambda \begin{pmatrix} R \\ F \end{pmatrix} \tag{4.13}$$

$$\qquad \mathbf{M} \qquad\quad \mathbf{v}_t \qquad\ \mathbf{v}_t$$

It is helpful to have the RHS of Eqn 4.13 in a matrix form similar to the LHS. To do this we employ the identity matrix, $\mathbf{I}$ (Box 4.1). Multiplying any matrix by the identity matrix leaves the matrix unchanged (therefore $\mathbf{M\,I} = \mathbf{M}$ on the LHS):

$$\begin{pmatrix} 0 & fs_{0,1} \\ s_{1,2} & 0 \end{pmatrix} \begin{pmatrix} R \\ F \end{pmatrix} = \lambda \begin{pmatrix} 1 & 0 \\ 0 & 1 \end{pmatrix} \begin{pmatrix} R \\ F \end{pmatrix}$$

$$\qquad \mathbf{M} \qquad\quad \mathbf{v}_t \quad\ \lambda \quad\ \mathbf{I} \quad\ \mathbf{v}_t$$

Now multiply the identity matrix $\mathbf{I}$ by the scalar λ:

$$\begin{pmatrix} 0 & fs_{0,1} \\ s_{1,2} & 0 \end{pmatrix} \begin{pmatrix} R \\ F \end{pmatrix} = \begin{pmatrix} \lambda & 0 \\ 0 & \lambda \end{pmatrix} \begin{pmatrix} R \\ F \end{pmatrix} \tag{4.14}$$

$$\qquad \mathbf{M} \qquad\quad \mathbf{v}_t \qquad \lambda\mathbf{I} \quad\ \mathbf{v}_t$$

We can now find a value for λ. Subtract the RHS from the LHS of Eqn 4.14:

$$\begin{pmatrix} 0 & fs_{0,1} \\ s_{1,2} & 0 \end{pmatrix} \begin{pmatrix} R \\ F \end{pmatrix} = \begin{pmatrix} \lambda & 0 \\ 0 & \lambda \end{pmatrix} \begin{pmatrix} R \\ F \end{pmatrix} = \begin{pmatrix} 0 \\ 0 \end{pmatrix}$$

$$\qquad \mathbf{M} \qquad\quad \mathbf{v}_t \qquad \lambda\mathbf{I} \quad\ \mathbf{v}_t$$

The LHS can be simplified by cancelling the common vector ($\mathbf{v}_t$) and subtracting the two square matrices:

$$\begin{pmatrix} 0-\lambda & fs_{0,1}-0 \\ s_{1,2}-0 & 0-\lambda \end{pmatrix} \begin{pmatrix} R \\ F \end{pmatrix} = \begin{pmatrix} 0 \\ 0 \end{pmatrix}$$

$$\qquad\quad \mathbf{M} - \lambda\mathbf{I} \qquad\quad \mathbf{v}_t$$

to give:

$$\begin{pmatrix} -\lambda & fs_{0,1} \\ s_{1,2} & -\lambda \end{pmatrix} \begin{pmatrix} R \\ F \end{pmatrix} = \begin{pmatrix} 0 \\ 0 \end{pmatrix} \tag{4.15}$$

$\quad\quad \mathbf{M} - \lambda \mathbf{I} \quad\quad \mathbf{v}_t$

If the matrix $\mathbf{M} - \lambda \mathbf{I}$ in Eqn 4.15 has an inverse (Box 4.1) then we could multiply both sides of the equation by the inverse matrix:

$$\begin{pmatrix} -\lambda & fs_{0,1} \\ s_{1,2} & -\lambda \end{pmatrix} \begin{pmatrix} a & b \\ c & d \end{pmatrix} \begin{pmatrix} R \\ F \end{pmatrix} = \begin{pmatrix} 0 \\ 0 \end{pmatrix} \begin{pmatrix} a & b \\ c & d \end{pmatrix}$$

$\quad \mathbf{M} - \lambda \mathbf{I} \quad$ inverse $\;\mathbf{v}_t$ of $\quad\quad$ inverse of
$\quad\quad\quad\quad\quad \mathbf{M} - \lambda \mathbf{I} \quad\quad\quad\quad \mathbf{M} - \lambda \mathbf{I}$

Multiplying the square matrix $\mathbf{M} - \lambda \mathbf{I}$ by its inverse on the LHS would give the identity matrix, $\mathbf{I}$ (by definition—see Box 4.1), whilst the RHS would reduce to 0:

$$\begin{pmatrix} 1 & 0 \\ 0 & 1 \end{pmatrix} \begin{pmatrix} R \\ F \end{pmatrix} = \begin{pmatrix} 0 \\ 0 \end{pmatrix}$$

$$\begin{pmatrix} R \\ F \end{pmatrix} = \begin{pmatrix} 0 \\ 0 \end{pmatrix}$$

This is unhelpful as we are left with the trivial solution that R and F are equal to 0. To overcome this problem we need to assume that the matrix $\mathbf{M} - \lambda \mathbf{I}$ does *not* have an inverse. This is true if the *determinant* (Box 4.1) *of the matrix is equal to 0*. This assumption can then be used to find a value for λ:

$$\begin{vmatrix} -\lambda & fs_{0,1} \\ s_{1,2} & -\lambda \end{vmatrix} = 0 \tag{4.16}$$

The determinant in Eqn 4.16 is referred to as the **characteristic determinant**. The whole equation 4.16 is called the **characteristic equation**. We can now evaluate the characteristic determinant and therefore solve the characteristic equation:

$$(-\lambda \times -\lambda) - fs_{0,1}s_{1,2} = 0$$

$$\lambda^2 = fs_{0,1}s_{1,2} \tag{4.17}$$

We are now left with a quadratic equation (4.17). Initially this poses a problem because a quadratic equation has two solutions (or roots); in other words, λ can have two values. But earlier we had assumed that the square matrix $\mathbf{M}$ could be replaced by a single value, λ. Effectively this becomes true as the largest of the two λ values, referred to as the **dominant root**, has most influence on the dynamics. Note that the dominant root may be complex or negative. A negative dominant root is biologically meaningless in this application (but see Chapters 5 and 7) whilst complex roots are discussed in Chapter 7. In mathematics the values of λ are called the **eigenvalues** and the corresponding values of R and F are the **eigenvectors**. The eigenvalues may also be referred to as the latent roots or the characteristic values of the matrix, **M**. Similarly, the eigenvectors are known as the latent or characteristic vectors. (In passing it is worth noting that in finding values for R and F we have found solutions for the Eqns 4.9 and 4.10. Matrix methods have a wide application in the solving of simultaneous equations.) Finally, it may be helpful to know that Eqns 4.13–4.16 can be written in a general mathematical shorthand for any size of matrix **M** and vector **v** (as Eqn 4.8):

$$\mathbf{Mv}_t = \lambda \mathbf{v}_t$$

$$\mathbf{Mv}_t - \lambda \mathbf{Iv}_t = 0$$

$$(\mathbf{M} - \lambda \mathbf{I})\mathbf{v}_t = 0$$

The requirement for the non-trivial solution is that

$$|\mathbf{M} - \lambda \mathbf{I}| = 0,$$

values of λ being found by solution of the characteristic equation.

To reinforce all these theoretical points let us consider a specific example. If $f = 100$, $s_{0,1} = 0.1$ and $s_{1,2} = 0.5$ then from Eqn 4.17:

$$\lambda^2 = 100 \ 0 \times 1 \ 0 \times 5$$

$$\lambda^2 = 5$$

$$\lambda = \pm \sqrt{5}$$

$+\sqrt{5}$ is both the larger value (and therefore the dominant root) and the one which is ecologically meaningful. We can now use this value of λ to produce a prediction of the rate of increase of R and F (based on Eqn 4.12):

$$\mathbf{v}_{t+1} = \sqrt{5} \ \mathbf{v}_t$$

or

$$\underset{\mathbf{v}_{t+1}}{\begin{pmatrix} R \\ F \end{pmatrix}} = \sqrt{5} \underset{\mathbf{v}_t}{\begin{pmatrix} R \\ F \end{pmatrix}}$$

It is important to note that the model predicts that both R and F increase at the same rate $\sqrt{5}$, and therefore predicts that they maintain the same ratio of R to F over time. This is equivalent to stating that a **stable age structure** is maintained with time. A quirk of this particular matrix model is that it produces oscillations from year to year (Fig. 4.3). The yearly increase by $\sqrt{5}$ (λ) therefore needs to be viewed over a two-year period, e.g. from year 4 to 6 the rosette numbers increased from 500 to 2500, i.e. a 5-fold increase, which is equivalent to a yearly increase of $\sqrt{5}$. Similarly, the stable age structure is seen only in even years (2, 4, 6 ...).

We can quantify the eigenvector and therefore determine the ratio of R to F as follows, using Eqn 4.13:

$$\begin{pmatrix} 0 & fs_{0,1} \\ s_{1,2} & 0 \end{pmatrix} \begin{pmatrix} R \\ F \end{pmatrix} = \sqrt{5} \begin{pmatrix} R \\ F \end{pmatrix}$$

Using the given values for f, $s_{0,1}$ and $s_{1,2}$ and multiplying out the LHS and RHS:

$$\begin{pmatrix} 10F \\ 0.5R \end{pmatrix} = \begin{pmatrix} \sqrt{5}\,R \\ \sqrt{5}\,F \end{pmatrix}$$

In effect we now have two equations: $10F = \sqrt{5}R$ and $0.5R = \sqrt{5}F$. These two equations are equivalent because rearrangement of either of the two equations produces $R = 2\sqrt{5}F$ or $R = 4.47F$.

We have now achieved both parts of the analysis described at the beginning of this section: we have found a value for λ, the finite rate of change, by determining the eigenvalue of the matrix and we have found the ratio of R to F by quantifying the eigenvector.

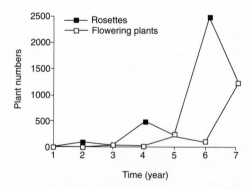

Fig. 4.3 Simulation of population dynamics of rosettes and flowering plants (Eqn 4.11) with values of $f = 100$, $s_{0,1} = 0.1$ and $s_{1,2} = 0.5$.

These techniques apply to more complex examples in which there are more than two ages, stages or sizes of organisms. The number of eigenvalues is equal to the number of rows or columns and therefore the number of ages, stages or sizes in the projection matrix **M**. Although the determination of eigenvalues becomes more difficult as the matrix increases in size (see Section 4.2.4), the principle continues to apply that it is the dominant eigenvalue that is important. Thus however big the projection matrix is, it can always be reduced to the dominant eigenvalue to describe the dynamics of the component stages of the population. Furthermore, the assumption of a stable age structure continues to hold, given by the values in the eigenvector.

Although we have focussed here on an age-structured population, it should be noted that many of the details and results of the model are also relevant to stage or size-structured populations.

Question 4.2

How could we check the validity of the model in the field?

4.1.5 Relationship between first- and second-order equations

As a slight digression, it is worth noting that the two original first-order equations $R_{t+1} = F_t f s_{0,1}$ and $F_{t+1} = R_t s_{1,2}$ (Eqns 4.9 and 4.10) could be combined to give a single second-order equation, in the same way that the discrete versions of the Lotka–Volterra equations were combined in Section 3.5.1 of Chapter 3. If the subscripts in the second equation are reduced by one time step on both sides to give $F_t = R_{t-1} s_{1,2}$ then we can substitute the RHS into the first equation:

$$R_{t+1} = R_{t-1} s_{1,2} f s_{0,1}$$

This new equation is a second-order difference equation relating the number of rosettes in year $t + 1$ (R_{t+1}) to those in year $t - 1$ (R_{t-1}), i.e. a difference of two years.

A similar second-order equation for flowering plants can be derived by reducing the time subscript in the first equation by 1 to give $R_t = F_{t-1} f s_{0,1}$ and substituting into the second equation giving $F_{t+1} = F_{t-1} f s_{0,1} s_{1,2}$.

These two second-order equations show that both rosettes and flowering plants increase at the same rate ($f s_{0,1} s_{1,2}$) over a two-year period. Using the previously assumed values of $f = 100$, $s_{0,1} = 0.1$ and $s_{1,2} = 0.5$, we see that:

$$R_{t+2} = 5R_t$$

$$F_{t+2} = 5F_t$$

Thus the rate of increase (R_{t+2}/R_t or F_{t+2}/F_t) over two years is 5, which is equivalent to the yearly rate ($\sqrt{5}$, i.e. λ) multiplied by itself (see Fig. 4.3). This therefore provides a useful check for our value of λ and an insight into the relationship between first- and second-order equations. The principle of converting two first-order equations to one second-order equation applies to higher-order equations.

4.2 Population dynamics of
the spear thistle *Cirsium vulgare*

In this section we apply the methods described above to a population of the plant *Cirsium vulgare* (spear thistle) in a sheep grazing experiment near Oxford. In particular we will see how simplification of a more complex model (Gillman *et al.* 1993) affects the prediction of the eigenvalues and how density dependence can be incorporated into this and similar models. The field data were collected by Silvertown and colleagues and are detailed in Bullock *et al.* (1994).

4.2.1 The experiment and measurement of the parameters

The sheep grazing experiment was set up in 1986 on species-poor grassland dominated by *Lolium perenne* and *Agrostis stolonifera*. The experiment is fully factorial (2 summer grazing levels × 2 winter grazing levels × 2 spring grazing levels) with two randomized blocks (Table 4.1). Each treatment replicate is applied to a 50 m × 50 m paddock. In summer the two grazing levels are applied by adjusting the stocking rate to produce either a 3 cm or 9 cm sward height.

Cirsium vulgare is a monocarpic perennial which is usually biennial (Fig. 4.4). The abundance and size-distribution of *C. vulgare* rosettes in each paddock have been monitored in April since 1986. Survival of various life-history stages and plant fecundity were also determined in each paddock. Seedling emergence usually occurs in early spring (January–April) and requires a gap in the vegetation. Three rosette size-classes were recognized in censuses: small rosettes (<10 cm) which

Table 4.1 Design of the grazing experiment at Little Wittenham Nature Reserve, near Oxford, England. A minus sign indicates no grazing during the relevant period. All paddocks were grazed in summer to either 3 cm or 9 cm in height.

Summer 3 cm			Summer 9 cm		
Treatment	Winter	Spring	Treatment	Winter	Spring
A	–	–	E	–	–
B	–	+	F	–	+
C	+	–	G	+	–
D	+	+	H	+	+

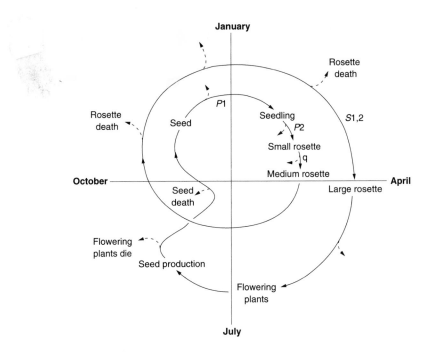

Fig. 4.4 Life-history of the spear thistle, *Cirsium vulgare* and detail of survival parameters.

die or become medium (10–20 cm) or large rosettes (>20 cm) by the next year. Only large rosettes were able to flower in the same year. Rosette size-class was a feature of the model of Gillman *et al.* but in this analysis we will treat all rosettes (R) as identical, assuming they are all medium-sized in the year of seedling emergence and large and therefore able to flower (F) in the following year. A second feature of the model of Gillman *et al.* was the inclusion of a seed bank in which seed may be dormant for one or more years. In this model we will ignore the role of the seed bank.

We will use Eqns 4.9 and 4.10:

$$R_{t+1} = f s_{0,1} F_t$$

$$F_{t+1} = s_{1,2} R_t$$

Explanation of the parameters and estimates for $s_{0,1} = p_1 p_2 q$ and $s_{1,2}$ for block 2 of the experiment are given in Table 4.2 and Fig. 4.4.

4.2.2 Estimation of eigenvalue and eigenvector

The values of λ (eigenvalue) from the simplified model are given in Table 4.2 and are compared with the results from the more complex model in Gillman *et al.* (with its extra components of rosette size-class and seed bank). λ for the simple

Table 4.2 Field parameter estimates from different grazing treatments (A–H, Table 4.1), following Bullock *et al.* (1994) and Gillman *et al.* (1993). m, maximum fraction of seeds able to germinate; p_2, probability of seedling survival from germination to small rosettes. Probabilities of survival of rosettes; q, small to medium rosettes; $s_{1,2}$, medium to large rosettes. f, average number of seed able to germinate per plant (fecundity $\times$ m). p_1, the probability of seed surviving before germination, was found to be unaffected by the treatments and taken as constant (0.24). λsimple was derived from Eqn 4.17 whilst λcomplex was derived from the simulation in Gillman *et al.* (1993). λfield was estimated from census data in three ways: as the geometric mean (g. mean) of all N_{t+1}/N_t; as the maximum N_{t+1}/N_t or as the intercept of $\ln(N_{t+1}/N_t)$ against N_t where N is $R + F$ (see Chapter 2 for details of the first and third methods). R/F is the ratio of rosettes to flowering plants produced either by the model or observed in the field. Treatment E is not included in the field data because in 5 out of 9 years the density was zero.

	Treatment							
	A	B	C	D	E	F	G	H
m	0.375	0.456	0.456	0.469	0.113	0.163	0.381	0.438
p_2	0.217	0.343	0.260	0.467	0.111	0.192	0.492	0.700
q	0.084	0.084	0.246	0.246	0.084	0.084	0.246	0.246
$s_{1,2}$	0.35	0.35	0.546	0.546	0.35	0.35	0.546	0.546
f	627	449	263	577	134	222	116	348
λsimple	0.98	1.04	1.49	2.95	0.32	0.55	1.36	2.80
λcomplex	1.34	1.48	1.25	2.48	0.77	0.96	1.52	2.36
λfield (g. mean)	1.52	1.01	1.19	1.26	—	0.78	0.63	0.93
λfield (max)	31.0	7.78	6.11	4.47	—	6.00	5.43	2.67
λfield (intercept)	9.46	1.67	1.97	1.87	—	1.82	1.34	2.26
R/F model	2.80	2.98	2.72	5.40	0.93	1.57	2.48	5.13
R/F field	5.04	14.27	6.30	23.3	—	2.0	2.32	5.18

model was obtained analytically as $+ \sqrt{(s_{0,1} \, s_{1,2} f)}$ (Eqn 4.17) whilst λ for the complex model was obtained by computer simulation over 50 generations. Apart from treatments A and B the results from the two models are quite similar. The predicted effect of grazing on λ is clear for both models. Winter and spring grazing combined (treatments D and H) give the highest values of λ. The effect of summer grazing is not so pronounced amongst these two treatments, with D very similar to H. Summer grazing is also relatively unimportant in treatments C and G. However, both treatments A and B are higher than E and F, respectively, suggesting that summer grazing is more important where winter grazing is less intense. The models are consistent in predicting that no winter grazing and mild summer grazing (treatments E and F) will result in λ values less than 1 and therefore a decreasing population. This is an important result for managers who do not want their fields infested with biennial 'weeds' such as *Cirsium vulgare*. The effects of grazing on λ are likely to be due to creation of gaps for germination (Silvertown & Smith 1989), alteration of intra- and interspecific competition and direct grazing of the thistle.

When we come to compare the λ values from the model with those from the field we face a problem—how can we measure λ? In Chapter 2 we noted that it is only appropriate to examine λ at low population densities, where density dependence is least important, yet in this experiment we are likely to have some density dependence as populations are not just geometrically increasing or decreasing (see Fig. 4.5 which shows the dynamics of *Cirsium* in the plots of block 2). Thus the geometric mean (g.m.) of λ (Table 4.2) is not very useful—indeed we might expect for a regulated population for that value to approach 1. There is some support for this assertion in Table 4.2; the plot g.m.s are distributed around 1, indeed the g.m. of the plot g.m.s is 1.007. The other two methods of determining λ for each treatment plot are to find the maximum value of λ (which we would expect to occur at small population sizes) or the intercept of the regression of $\ln(N_{t+1}/N_t)$ against N_t (Chapter 2). Both of these measures are problematic owing to the stochastic effect of small population size. Thus, when N_t is small, e.g. just one individual, an extra few plants in the following year will cause N_{t+1}/N_t (and therefore the intercept or maximum λ) to be high. A few extra plants when there are already (say) 100 will have little effect on N_{t+1}/N_t. Therefore small random events such as the opportunistic colon- ization of a few plants will have a biased (the population cannot drop below 0) and a disproportionate effect on estimation of λ when N_t is small. This is the reason why density dependence is often detected from a series of random numbers using the regression technique (Chapter 2): the population can only increase (and perhaps by large amounts) at small population sizes whereas decrease is more likely at large population sizes. These considerations help to explain why the deviations between λfield and λmodel are sometimes so large in Table 4.2. However, when

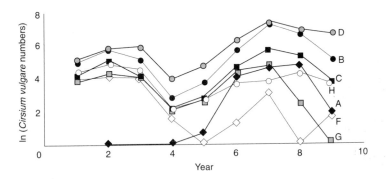

Fig. 4.5 Dynamics of *Cirsium vulgare* in plots of Block 2. The plots receive the grazing treatments in Table 4.1.

the mean population size is relatively large then the agreement between the intercept method and the simple model is reasonably good. This is shown by comparing λfield (intercept) divided by λsimple with the means as shown in Table 4.3.

Table 4.3 Assessment of the similarity of λfield (intercept) and λmodel with change in mean population size.

	A	B	C	D	F	G	H
Mean population size	35.1	375	111	569	22	44	61
λfield (intercept)/λsimple	9.65	1.60	1.32	0.63	3.31	0.99	0.81

The two highest discrepancies between λsimple and λfield intercept occur for treatments A and F, the two treatments with lowest mean values.

It should be noted that for λ and for R/F (below) the survival and fecundity parameters were not estimated from the census data and therefore provide an independent test of λfield and R/Ffield.

The structure of the population can be expressed as the ratio of rosettes to flowering plants. The expected value of R/F is given by two equations from Eqn 4.13, for example:

$$s_{1,2}R = \lambda F$$

Therefore $R/F = \lambda/s_{1,2}$

This gives the predicted ratio in Table 4.2 (R/Fmodel). These values vary from 0.93 to 5.40, i.e. we expect anything from the same number to five times as many rosettes as flowering plants in the field. From the censuses the observed values, which were not used in the construction of the model, can be taken as (small + medium)/large rosettes, i.e. the plants which are likely to be derived from seed in that spring divided by those which are likely to flower in that year. These observed values vary from year to year so the data in Table 4.2 are derived from averages of 1993–1995. The observed range is from 2 to more than 20, much higher than expected from the model, although the agreement for treatments F, G and H is good. However, when we look at the fraction of small/(medium + large) this gives the following, as shown in Table 4.4.

Table 4.4 The ratio of small to medium/large rosettes. There were no rosettes in E in two of the years.

A	B	C	D	E	F	G	H
1.51	2.14	3.95	4.03	—	0.35	1.31	5.83

These values are much closer overall to those predicted from the simple model.

This suggests that the mismatch between observed and predicted R/F may lie in the categorization of the rosettes and that the model needs some description of rosette size structure.

In summary, the simple model has proved useful in determining the value of λ, which can be related to the grazing treatments and the value of λ calculated from independent census data. The model has also highlighted potential problems in the assessment of $R : F$. Only with such comparisons between models and field can we move towards the ideal of the simplest realistic model for population dynamics.

Question 4.3

What would be the effect on the model results (eigenvalue and eigenvector) of assuming rosettes were small rather than medium in the first year (Fig. 4.4)?

4.2.3 Incorporation of density dependence and analysis of its effects on population dynamics

Density dependence may be operating at one or more points in the life-cycle. In constructing models of structured populations with density dependence we have a series of options. For example, with *Cirsium vulgare*, density dependence may be operating at the seed stage owing to a limited number of microsites for germination. If only one seed can germinate per microsite and the density of microsites is fixed, then an increased density of seeds will produce a decreased fraction of germination, i.e. a density-dependent mechanism. Thus the fraction of seed germinating in year t (g_t) may be determined mainly by the proportion of bare ground, which is in turn dependent on grazing (Silvertown & Smith 1989). It can also be envisaged that g is a function of the number of all *C. vulgare* rosettes; thus a density-dependent mechanism can be envisaged, operating with intensity a, in which rosettes reduce the fraction of ground available for germination, and therefore the fraction of seeds able to germinate, from a maximum of m (Table 4.2). A negative exponential relationship can be assumed for the density dependence (Gillman *et al.* 1993), as in Chapter 2:

$$g_t = m\,e^{-aR_t} \tag{4.18}$$

In Chapter 6 the effect of gap dependence on population dynamics is considered.

There may also be intraspecific competition between rosettes and so the fraction of surviving rosettes or the yield and therefore fecundity of flowering plants may be dependent on rosette density. Watkinson (1980, 1987) reviewed examples of yield–density relationships and formulated a new model to describe the population dynamics of annual plants using the scramble–contest model of Hassell (1975) previously applied to insect populations (see Chapter 2):

$$N_{t+1} = \frac{\lambda N_t}{(1 + aN_t)^b + w\lambda N_t}$$

where a and b are the parameters of the Hassell model, w is the degree of self-thinning and λ is the finite rate of population change. This model could be extended to longer lived plants where there is intraspecific competition at one stage.

The type of dynamics produced depends on the complexity of the age structure, the point(s) at which density dependence is incorporated, the form of the density dependence and (as seen in Chapter 2) the finite rate of increase. In Gillman *et al.* using Eqn 4.18 and other parameter values from treatments D and H produced two-point limit cycles for *Cirsium vulgare* (Fig. 4.6b). These were most pronounced at the small and medium rosette stages. For example, the small rosettes fluctuated almost four-fold in numbers in treatment D, compared with virtually no change amongst the large rosettes (Fig. 4.6b). Two other treatment conditions (C and G) resulted in a stable equilibrium in both blocks (treatment C, Fig. 4.6a).

There was a significant positive correlation between the observed mean rosette numbers (per treatment replicate) and the predicted equilibrium values from the model (the mean of the two-point limit cycle values was used). As noted for the comparison of λ and R/F between model and field, this was a reasonable test of the model as the model values were not derived from the census data.

De Kroon *et al.* (1987) undertook similar analyses to those with *Cirsium vulgare* for populations of the perennial rosette-forming herb *Hypochaeris radicata*. In their study, mowing was the management regime and density dependence was incorporated at both germination and seedling establishment, which were in turn functions of gaps in the vegetation. They used a stage-structured model with four stages (seeds and three stages of rosette, some of which were further divided by size or stage; Fig. 4.7a). The time steps were between seasons rather than years. De Kroon *et al.* also argued that a sigmoidal (s-shaped) rather than negative exponential form of density dependence was most appropriate. The results of their model under three mowing frequencies are shown in Fig. 4.7(b). Increasing mowing frequency produced more gaps in the vegetation and higher germination, seedling establishment/survival and rosette survival. The net result was that population growth rates were predicted to be higher with increased mowing frequency, with the highest mowing frequency producing damped oscillations in the model (Fig. 4.7b).

Solbrig *et al.* (1988), in a model of the dynamics of *Viola fimbriatula*, assumed that seedling survival was a linear function of adult density, i.e. a discrete logistic density-dependent function (Chapter 2). This produced stable oscillations under certain model conditions (Fig. 4.8) due in the authors' words to the 'combined effect of density dependence of seedling survival and the one oscillation cycle lag in incorporating surviving seedlings into the adult population'. They also tried other

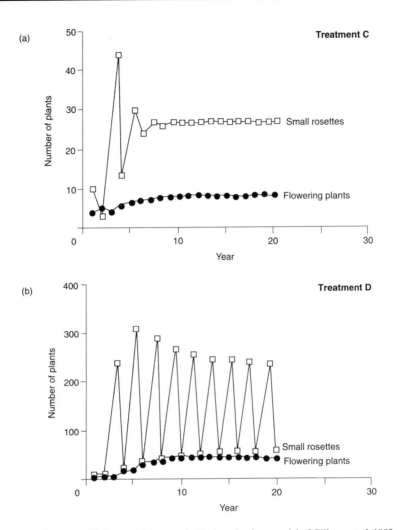

Fig. 4.6 (a) Stable equilibrium and (b) two-point limit cycles from model of Gillman *et al.* 1993

forms of density dependence, such as the negative exponential, and noted how the form of the density dependence could affect the dynamics.

The application of density dependence to structured population models, particularly with field-estimated parameter values, is still in its infancy (see Chapter 9 in Caswell 1989). The *Cirsium vulgare* and *Hypochaeris radicata* studies are rare examples, highlighting the value of field experiments in both the estimation of model parameter values and the testing of models.

The techniques discussed for *Cirsium vulgare* are now reinforced by applying them to a bird population. Two developments of the model will be apparent. First, the population is described by an explicitly stage-structured model and, second,

(a)

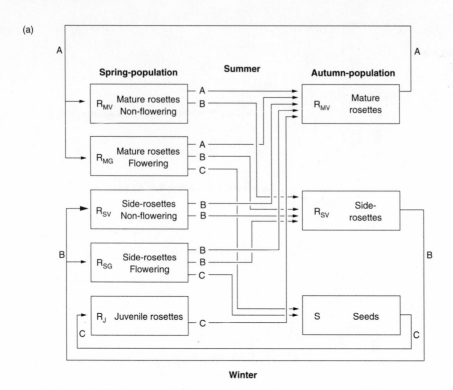

(b)

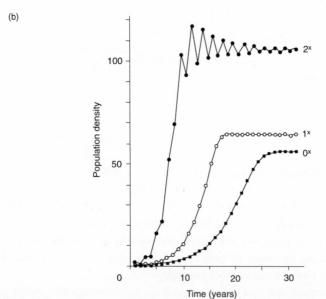

Fig. 4.7 (a) Life-history diagram of *Hypochaeris radicata* (de Kroon *et al.* 1987). The three main life-history pathways are survival of adults (A), vegetative ramification (B) and sexual reproduction (C). The diagram shows summer and winter processes separately. (b) Simulated population growth of *H. radicata* with three mowing frequencies.

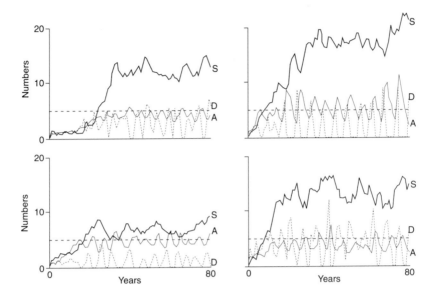

Fig. 4.8 Density-dependent dynamics of *Viola fimbriatula*. Four simulations over 80 years are shown. The numbers of seeds in the seed pool (*S*), seedlings (*D*) and adults (*A*). The scale for the ordinate is in hundreds for seedlings and adults, but in thousands for the seed pool. The dashed line at 500 adults is the density beyond which no seedlings survive.

there are three rather than two stages. We will also be dealing, in comparison with *Cirsium vulgare*, with a sparse set of data typical of many species for which conservation action is required.

4.3 Three-stage model of Cabot's tragopan

Cabot's tragopan *Tragopan caboti* is endemic to the lower montane zone of south-eastern China, where it inhabits evergreen deciduous forest and mixed deciduous-coniferous forest at an altitude of 800–1400 m (Collar & Andrew 1988).

Cabot's tragopan breeds once a year, starting in late March or early April and ending in May or June. The following details are from Wuyanling (and Philip McGowan *pers. comm.*, Zhang Ding Chang-qing and Zheng, 1990), where the sex ratio of adult male to females is approximately 1 : 1.

Nesting failure is very high. During 1984–1993, 33 nests were found in the wild and only five (i.e. about 15%) of them hatched successfully. The reason for nest loss was attributed to various factors such as predation by jays and egg-collection by local people. Completed clutches contain two to six eggs, with an average of 3.5. Because of the bad winter weather (snow and heavy rain) and the predators, the survival rate of chicks is generally not more than 50%. The ratio of adults to one-year-old birds is 4.6:1. In the wild the bird starts breeding in the third year. Life expectancy in the wild is unknown; in captivity the longest recorded is 8 years.

Using these data the finite rate of population change can be estimated. The Cabot's tragopan population at Wuyanling can be divided conveniently into three stages: egg (*E*, age 0), juvenile (*J*, age 1 and 2 years), adult (*A*, age 3–8 years). At present we only have limited data from which to estimate stage-dependent survival and fecundity (above):

egg to juvenile survival, 0.15

juvenile survival, 0.5

adult survival rates from year to year are unknown but we assume high values relative to juvenile survival, such as 0.9

average fecundity is 3.5

This allows us to construct the following stage-structured model:

$$\begin{pmatrix} E \\ J \\ A \end{pmatrix} = \begin{pmatrix} 0 & 0 & 3.5 \\ 0.15 & 0.5 & 0 \\ 0 & 0.9 & 0.9 \end{pmatrix} \begin{pmatrix} E \\ J \\ A \end{pmatrix}$$

$$\mathbf{v}_{t+1} \qquad \mathbf{M} \qquad \mathbf{v}_t$$

From the model we can estimate the maximum rate of increase (the dominant eigenvalue λ of the matrix $\mathbf{M}$) as 1.33 by solution of the characteristic equation (Box 4.2 gives details of the determination of the characteristic equation for three or more stages):

$$-\lambda^3 + 1.4\lambda^2 - 0.45\lambda + 0.4725 = 0$$

This has a corresponding stage structure of 1 egg : 0.18 juveniles : 0.38 adults, evaluated from $\mathbf{Mv}_t = \lambda \mathbf{v}_t$. If the juvenile stage is divided by two, this gives a predicted yearling to adult ratio of 0.09 : 0.38 or 1 : 4.2 which is very close to the observed value of 1 : 4.6.

Table 4.5 Estimated number of birds present in the breeding season in Wuyanling National Natural Reserve, Zhejiang Province, south-eastern China.

Year	1985	1986	1987	1988	1989	1990	1991	1992
Number of birds	92	119	–	80	–	255	102	74

In Wuyanling numbers of birds have been estimated in 6 out of 8 years in the breeding season (Table 4.5).

The two missing years (1987 and 1989) mean that we can only use three pairs of years (1985, 1986; 1990, 1991; 1991, 1992) to generate three values of λ:

119/92; 102/255; 74/102, i.e. 1.29; 0.40; 0.73. Notice that the highest value, 1.29, agrees with the independent estimate from the stage-structured model of 1.33. Therefore despite the lack of data, there is a surprisingly good agreement between the predictions of the model and the independent census predictions.

Box 4.2 Determinant and characteristic equation of a 3 × 3 matrix

The determinant of a 3 × 3 matrix is found as follows:

1 Construct an array of signs (+ or − values) across the matrix:

$$\begin{pmatrix} + & - & + \\ - & + & - \\ + & - & + \end{pmatrix}$$

2 Exclude the first row and first column of the matrix and calculate the determinant of the remaining 2 × 2 matrix (this determinant is known as the minor of the first row, first column element). For example with the Cabot's tragopan matrix in the text:

$$\begin{pmatrix} x & x & x \\ x & 0.5 & 0 \\ x & 0.9 & 0.9 \end{pmatrix}$$

Determinant equals $0.5 \times 0.9 - 0 \times 0.9 = 0.45$

3 Multiply the determinant of the minor by its corresponding element and +1 (its sign from **1** above).

4 Repeat procedures **2** and **3** for the first row, second column element and minor and the first row, third column element and minor.

5 Add the three values together to give the determinant of the 3 × 3 matrix. For example, with the Cabot's tragopan matrix:

$$\begin{vmatrix} 0 & 0 & 3.5 \\ 0.15 & 0.5 & 0 \\ 0 & 0.9 & 0.9 \end{vmatrix} = +0 \begin{vmatrix} 0.5 & 0 \\ 0.9 & 0.9 \end{vmatrix} - 0 \begin{vmatrix} 0.15 & 0 \\ 0 & 0.9 \end{vmatrix} + 3.5 \begin{vmatrix} 0.15 & 0.5 \\ 0 & 0.9 \end{vmatrix}$$

$$= 3.5(0.15 \times 0.9)$$

$$= 0.4725$$

To find the characteristic equation, follow the method for the 2 × 2 matrix. For example, with Cabot's tragopan:

Continued on page 108

Box 4.2 (*continued*)

$$\left| \begin{pmatrix} 0 & 0 & 3.5 \\ 0.15 & 0.5 & 0 \\ 0 & 0.9 & 0.9 \end{pmatrix} - \lambda I \right| = 0$$

$$\left| \begin{pmatrix} 0 & 0 & 3.5 \\ 0.15 & 0.5 & 0 \\ 0 & 0.9 & 0.9 \end{pmatrix} - \begin{pmatrix} \lambda & 0 & 0 \\ 0 & \lambda & 0 \\ 0 & 0 & \lambda \end{pmatrix} \right| = 0$$

$$\begin{vmatrix} -\lambda & 0 & 3.5 \\ 0.15 & 0.5 - \lambda & 0 \\ 0 & 0.9 & 0.9 - \lambda \end{vmatrix} = 0$$

Now follow the method for evaluating the determinant:

$$-\lambda \begin{vmatrix} 0.5 - \lambda & 0 \\ 0.9 & 0.9 - \lambda \end{vmatrix} - 0 \begin{vmatrix} 0.15 & 0 \\ 0 & 0.9 - \lambda \end{vmatrix} + 3.5 \begin{vmatrix} 0.15 & 0.5 - \lambda \\ 0 & 0.9 \end{vmatrix}$$

$$-\lambda(0.5 - \lambda)(0.9 - \lambda) + 3.5(0.15 \times 0.9)$$

$$-\lambda(0.45 - 1.4\lambda + \lambda^2) + 0.4725$$

to give the characteristic equation described in the text:

$$-\lambda^3 + 1.4\lambda^2 - 0.45\lambda + 0.4725 = 0$$

4.4 Sustainable harvesting of structured populations

In Chapter 3 we discussed how populations with continuous reproduction might be harvested. In the final section of this chapter we return to the theme of harvesting and ask how structured populations, described by matrices, might be affected.

4.4.1 A harvesting matrix

We begin with an unstructured population in discrete time in which the population dynamics are described by λ, the finite rate of population increase, which, as we have seen, is equivalent to the dominant eigenvalue of the population projection matrix for a structured population. We know that a population with $\lambda < 1$ will decline in numbers, so if a harvesting policy is to be sustainable it should not decrease the value of λ below 1. It is therefore possible to arrive at a simple definition of the maximum amount of a population which can be harvested. The fraction of the population surviving harvesting multiplied by the finite rate of increase must equal 1 (or more). Thus (1 – fraction harvested) $\lambda = 1$, so the maximum fraction harvestable is $(\lambda - 1)/\lambda$.

If $\lambda = 3$, the maximum fraction of the population which could be harvested is 2/3, i.e. $(\lambda - 1)/\lambda$. This assumes that λ is constant from year to year and that a constant fraction is removed in each time period.

Now consider the implications of age, size or stage structure. To explore the effect of population structure on harvesting consider the two-stage model of biennial plants used previously (Eqn 4.11):

$$\begin{pmatrix} R \\ F \end{pmatrix} = \begin{pmatrix} 0 & fs_{0,1} \\ s_{1,2} & 0 \end{pmatrix} \begin{pmatrix} R \\ F \end{pmatrix}$$

$$\mathbf{v}_{t+1} \qquad \mathbf{M} \qquad \mathbf{v}_t$$

The characteristic equation was $\lambda^2 = fs_{0,1}s_{1,2}$ (Eqn 4.17). We saw above that the maximum fraction harvested is given by $(\lambda - 1)/\lambda$, so it becomes of interest to see how manipulation of f, $s_{0,1}$ and $s_{1,2}$ affects this maximum fraction.

Assume that a fraction m_1 of flowering plants is harvested prior to setting seed and therefore the fraction of surviving plants is represented by the fraction $(1 - m_1)$. (Removal of flowering plants after seed set for a monocarpic species is not going to affect the population dynamics.) This harvesting mortality can be incorporated into the model as follows:

$$\begin{pmatrix} R \\ F \end{pmatrix} = \begin{pmatrix} 0 & fs_{0,1} \\ s_{1,2}(1 - m_1) & 0 \end{pmatrix} \begin{pmatrix} R \\ F \end{pmatrix}$$

$$\mathbf{v}_{t+1} \qquad \begin{array}{c}\mathbf{M} \text{ with harvesting} \\ \text{of } F\end{array} \qquad \mathbf{v}_t$$

In a similar way, we might imagine that a fraction m_2 of rosette plants is harvested. (Of course either m_1 or m_2 or both can have zero values.) This can also be included in the model:

$$\begin{pmatrix} R \\ F \end{pmatrix} = \begin{pmatrix} 0 & fs_{0,1}(1 - m_2) \\ s_{1,2}(1 - m_1) & 0 \end{pmatrix} \begin{pmatrix} R \\ F \end{pmatrix} \qquad (4.19)$$

$$\mathbf{v}_{t+1} \qquad \begin{array}{c}\mathbf{M} \text{ with harvesting} \\ \text{of } F \text{ and } R\end{array} \qquad \mathbf{v}_t$$

We can now determine λ for the new model with the harvesting mortalities (refer back to Eqns 4.16 and 4.17 if necessary).

The characteristic equation of 4.19 is

$$\lambda^2 = fs_{0,1}(1 - m_2)s_{1,2}(1 - m_1)$$ (4.20)

This is obviously similar to the characteristic equation without harvesting (Eqn 4.17). Indeed, if we define λ_u as λ when the population is unharvested and λ_h as λ when the population is harvested, then we can derive an interesting result from Eqn 4.20.

Taking the square root of Eqn 4.20:

$$\lambda_h = \pm\sqrt{(fs_{0,1}s_{1,2})\,[(1 - m_2)(1 - m_1)]}$$

As

$$\lambda_u = \pm\sqrt{(fs_{0,1}s_{1,2})}$$

$$\lambda_h = \pm\lambda_u\sqrt{[(1 - m_2)(1 - m_1)]}$$

Thus the value for λ_h is given by the original λ_u multiplied by the square root of $(1 - m_2)(1 - m_1)$. Using matrices we can approach this problem from a slightly different angle to shed more light on the expression $\sqrt{[(1 - m_2)(1 - m_1)]}$.

Let us decompose the square matrix from Eqn 4.19:

$$\begin{pmatrix} 0 & fs_{0,1}(1 - m_2) \\ s_{1,2}(1 - m_1) & 0 \end{pmatrix}$$

to give two square matrices multiplied together:

$$\begin{pmatrix} 0 & fs_{0,1} \\ s_{1,2} & 0 \end{pmatrix}\begin{pmatrix} 1 - m_1 & 0 \\ 0 & 1 - m_2 \end{pmatrix}$$

The left-hand matrix is the original transition matrix for the unharvested population (Eqn 4.11) and has an eigenvalue of λ_u. The right-hand matrix is composed of the two harvesting 'survivals' and can be referred to as a harvesting matrix (Lefkovitch 1967). We could use this method for any structured population to explore the effects of harvesting of different ages, sizes or stages on population dynamics.

Usher (1973) described how these results allow us to understand and explore the way that structured populations can be harvested and managed. He considered their application to Pacific sardine, various fish species in the North Sea, two whale species, red deer and laboratory populations of beetles, flies, *Daphnia* and *Collembola*. Matrix models have also been used to understand how to best conserve populations. For example, Manly (1990) gives the example of the work of Crouse *et al.* on loggerhead turtles, which showed that promoting juvenile survival rather than eggs in nests was more likely to halt population decline. Olmstead and Alvarez

Buyalla (1995) have explored the possibility of sustainable harvesting of two tropical palm species using matrix models. They calculated the population growth rates from stage-structured models and estimated the amount of adult trees which could be harvested per unit area. They concluded that just one species could be harvested, the other having a λ of only 1.05. The techniques of sensitivity and elasticity analysis (described in Caswell 1989 and de Kroon *et al.* 1986) can help in these studies by identifying the most important components of life-history in terms of their contribution to λ.

4.5 Conclusions

Matrices allow constructions of models of population dynamics for species with age, stage or size structure. This chapter illustrates the relationship of these models, through the dominant eigenvalue, to unstructured density-independent models (Chapter 2). Matrix models are seen to be valuable when analysing the effects of harvesting and other management such as grazing on species with complex life-histories. However, just as with the harvesting examples using the logistic equation in Chapter 3, we need to be cautious because these models are deterministic. Variations from year-to-year and site-to-site may alter λ or reduce the maximum sustainable yield well below its theoretical value—indeed stochasticity may render the latter term invalid for many species (see the discussion in Beddington 1979 and references therein).

CHAPTER 5

Dynamics of ecological communities

In Chapters 2–4 the emphasis has been on population dynamics of a single species or two interacting species such as predator and prey in a given locality. The latter may constitute a simple ecological community defined as all individuals of interacting species in a given area. Often, for convenience of sampling, the definition is relaxed to include all individuals in a given area. To ease the problem of sampling and retain the potential interactions between species, ecologists make use of guilds, defined as a set of species sharing a common resource such as a host-plant (Root 1967). Consideration of guilds has the advantage that interactions such as inter-specific competition are likely to be occurring between those species. Therefore, a community might comprise the guild of all herbivorous insects feeding on a species of plant, the plant itself and the natural enemies of the herbivores.

Even in the simplest community we are likely to be dealing with more than two species in which case the predator–prey models of Chapter 3 do not provide appropriate descriptions, so we need to develop models to deal with more than two species. In all these cases we will be interested in both the dynamics of the individual species and the dynamics, stability and structure of the whole community. The latter may be expressed in terms of changes in species richness or dominance over time and investigated through the response of the community to perturbation, for example, by removal of a species.

The modelling of community dynamics has three related problems which will be tackled in this chapter. First, there is a need to consider the full range of interactions between species, described by all combinations of 0/+/−, where 0 represents no interaction, +, a positive (beneficial) effect of species i on j and −, a detrimental effect of i on j (Fig. 5.1). The second problem is that we need to assign a strength to these interactions. In this chapter we describe how interactions such as interspecific competition (− −) and mutualism (+ +) can be modelled in a similar way to predator–prey or host–pathogen interactions (− +), and that this can be generalized as a community matrix in which both the sign and magnitude (i.e. strength) of the interaction can be incorporated. We will use matrix methods to investigate the stability and dynamics of communities. The final problem is the sheer number of species which may be involved. In response to this we will see how species can be objectively removed from the model, moving towards the aim of the simplest realistic model outlined in Chapter 1.

The above considerations of communities generally assume a pool of species within a given area, each of which can change in abundance over time (even become

		Effect of species j on i		
		+	0	-
Effect of species i on j	+	++	+0	+-
	0	0+	00	0-
	-	-+	-0	--

Fig. 5.1 Classification of interactions.

locally extinct) but do not assume any immigration and possible replacement of species as seen during successional change. The final section of the chapter will consider ecological models which address succession and again use matrix-based analyses.

5.1 Models of interspecific competition

In Chapter 2 we considered the possibility of *intra*specific competition as a mechanism producing density dependence. In this chapter our attention is on competition between species, i.e. *inter*specific competition. This will serve as a prelude to a generalized description of interactions between species.

5.1.1 Competition experiments and derivation of competition coefficients

Much of the classic work on interspecific competition has involved simple microcosms such as insects in stored grain (Crombie 1945, 1946, 1947; Park *et al.* 1964), aquatic protozoa (Gause 1934, 1935) and yeast (Gause 1932, 1934). For example, the experiments of Crombie (1945, 1946) showed how three combinations of the beetles *Tribolium confusum* and *Oryzaephilus surinamensis* converged to the same population equilibrium. The beetles were fed on wheat, presented as either cracked grain or flour. In cracked wheat each species cultured in isolation increased to between 420–450 adults in 150 days (represented as carrying capacities K_1 and K_2 in Table 5.1). The change in numbers over time is plotted on the phase-plane in Fig. 5.2. However, when the species were combined, *Tribolium* reached 360 individuals and *Oryzaephilus* 150 individuals; thus the total was greater than the species in monoculture (represented as N_1^* and N_2^* in Table 5.1). The results were independent of the initial number of beetles indicating that this was a (globally) stable equilibrium. As Pontin (1982) noted, 'the total number of beetles in mixed culture at equilibrium is equal to or greater than the carrying capacity number of either (species) alone so the combination may be more efficient at converting grain to beetles'. Changing the medium from flour grains to fine flour resulted in the extinction of *Oryzaephilus*. Small pieces of tubing which provided shelter for *Oryzaephilus* allowed the latter to survive under fine flour conditions (Table 5.1).

Table 5.1 Values of carrying capacity of *Tribolium confusum* (species 1) and *Oryzaephilus surinamensis* (species 2) in isolation (K_1 and K_2) and equilibrium in mixtures (N_1^* and N_2^*) estimated from experiments. Values for the competition coefficients (β_{12}, β_{21}) are derived in text. (Data in Pontin 1982 from Crombie 1946.)

	Equilibrium numbers				Competition coefficients	
	Alone		Together			
	K_1	K_2	N_1^*	N_2^*	β_{12}	β_{21}
Cracked wheat	425	445	360	150	0.4	0.8
Fine flour 1 mm tubes	175	400	175	80	Small	1.8

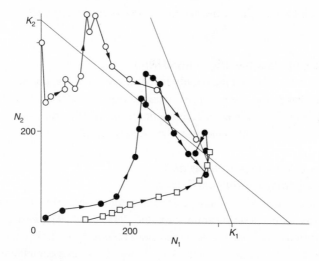

Fig. 5.2 Population trajectories from three competition experiments started with different numbers of *Tribolium* (N_1) and *Oryzaephilus* (N_2) in cracked wheat cultures. The lines represent zero growth of each species—see text and Fig. 5.3. (From Crombie 1946 reprinted in Pontin 1982.)

The same general effects of habitat structure on coexistence were found in the predator–prey experiments of Huffaker (1958, see Chapter 6).

Tribolium larvae are cannabalistic—the larvae and adults eat eggs and pupae of their own species and also those of *Oryzaephilus*. Adult *Oryzaephilus* also destroys *Tribolium* eggs but at a lower rate than its own are destroyed by *Tribolium*. Therefore this interaction is part predation and part competition for a resource. This emphasizes the need for a generalized model in which a variety of interactions are encompassed.

A pair of competing species such as *Tribolium* and *Oryzaephilus* can be described by a pair of simultaneous non-linear differential equations similar to those used for the predator–prey interactions by Lotka and Volterra (Chapter 3) and, indeed, were developed by Volterra (1926, 1931) and Gause (1934):

$$\frac{dN_1}{dt} = \frac{r_1 N_1 (K_1 - N_1 - \beta_{12} N_2)}{K_1} \tag{5.1}$$

$$\frac{dN_2}{dt} = \frac{r_2 N_2 (K_2 - N_2 - \beta_{21} N_1)}{K_2} \tag{5.2}$$

where N_1 and N_2 are the densities of the competing species. β_{12} describes the fraction of species 2 converted into species 1. For example, if $\beta_{12} = 0.5$ then one individual of species 2 is equivalent to half an individual of species 1 in competition with species 1. β is known as the **competition coefficient** and is generalized to interactions between species i and j as β_{ij}. (The reason for using β will become apparent later.) r is the intrinsic rate of increase for a given species (Chapter 3) and K_1 and K_2 are the carrying capacities of species 1 and 2 in isolation. Note that Eqn 5.1 could also be written as:

$$\frac{dN_1}{dt} = \frac{r_1 N_1 [K_1 - (N_1 - \beta_{12} N_2)]}{K_1}$$

As β_{12} is the fraction of species 1 converted to species 2 then $N_1 + \beta_{12} N_2$ is effectively the density of N_1, replacing N_1 in the ordinary logistic equation (Chapter 3).

Investigation of stability with the phase-plane begins, as in Chapter 3, with the zero growth isoclines ($dN_1/dt = 0$ and $dN_2/dt = 0$). We will go through this more rapidly than in Chapter 3 as many of the principles are the same. If we take the example of *Tribolium* and *Oryzaephilus*, we find the values of K_1, K_2 (the carrying capacities of the two species in isolation), N_1^* and N_2^* (the equilibrium densities in mixtures) from the experiment (Table 5.1).

With zero growth, Eqns 5.1 and 5.2 become:

$$0 = \frac{r_1 N_1^* (K_1 - N_1^* - \beta_{12} N_2^*)}{K_1} \tag{5.3}$$

$$0 = \frac{r_2 N_2^* (K_2 - N_2^* - \beta_{21} N_1^*)}{K_2} \tag{5.4}$$

From Eqn 5.3 we have either $r_1 N_1^* = 0$ (the trivial solution) or $(K_1 - N_1^* - \beta_{12} N_2^*)/K_1 = 0$ which gives $N_1^* = K_1 - \beta_{12} N_2^*$. Similarly Eqn 5.4 yields $N_2^* = K_2 - \beta_{21} N_1^*$. As N_1^*, N_2^*, K_1 and K_2 are known (Table 5.1) we can use these equations to find β_{12} and β_{21} and plot the zero growth lines (Figs 5.2 and 5.3).

The outcome of competition can then be investigated on the phase-plane in the same way as for predator–prey interactions in Chapter 3. These outcomes are: (i) stable competition in which both species coexist, (ii) unstable competition in which one species always displaces the second species, i.e. there is a fixed hierarchy of

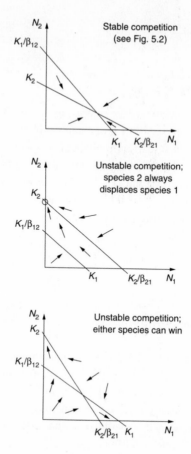

Fig. 5.3 Three outcomes of interspecific competition based on three combinations of zero growth isoclines from Eqns 5.3 and 5.4. The approximate direction of change in species 1 and 2 is indicated by the vectors.

competition and (iii) unstable competition in which either species can win (Fig. 5.3). (The assessment of competition coefficients in the field is considered in the context of generalized Lotka–Volterra models in Section 5.2.4.)

Other competition models have been developed which use similar equations to Eqns 5.1 and 5.2 above. For example, Hassell and Comins (1976) developed a discrete equation version:

$$N_{1,t+1} = \lambda_1 N_{1,t} [1 + a_1(N_{1,t} + \beta_1 N_{2,t})]^{-b1}$$

$$N_{2,t+1} = \lambda_2 N_{2,t} [1 + a_2(N_{2,t} + \beta_2 N_{1,t})]^{-b2}$$

This used four parameters compared with the three in Eqns 5.1 and 5.2. β_1 and β_2 are the competition coefficients; a_1 and a_2 give the threshold densities at which density dependence begins and b_1 and b_2 are parameters which describe the different types of intraspecific competition, with extremes of scramble and contest (Hassell 1975; see Chapter 2 for discussion of scramble and contest and description of Hassell's model). λ is the finite rate of population change. Atkinson and Shorrocks (1981) used the model of Hassell and Comins to explore the effect of aggregation of competing *Drosophila* species on coexistence. The degree of aggregation was modelled using the **negative binomial** (Box 6.1). An increased aggregation of the superior | competitor | promoted | coexistence | because | of | increased | intraspecific interactions. The role of aggregation is developed with respect to host–parasitoid models in Chapter 6.

Question 5.1

How do Eqns 5.1 and 5.2 differ from those used to describe predator–prey interactions (Eqns 3.11 and 3.12)?

5.2 The community matrix

5.2.1 Introduction
The **community matrix** allows insights into the stability of the community, with the dynamics of each species being described by a non-linear first-order differential equation. The original community matrix was devised by Levins (1968) as a matrix of Lotka–Volterra competition coefficients to describe community structure and predict community dynamics. The theory was developed by May (1972, 1973a) giving a general version of the community matrix (sometimes referred to as the **stability matrix**) which expressed the effect of species j on species i near equilibrium. It is this matrix which will be referred to in the following sections.

5.2.2 Construction and stability of the community matrix
The aim is to create a matrix of all interactions between the species in the community. It is assumed that species will be regulated so that, in the absence of any interspecies interactions, they will return to equilibrium. The following notation will be used:

α_{ij}: the interaction coefficients between species i and j (expressed as the effect of species j on the growth rate of species i); this includes α_{ii} the intraspecific interaction. (The magnitude of α ranges from 1 to 0, which represents no interaction. The sign of α is considered below.)

N_i: density or biomass of species i at time t.

r_i: the intrinsic rate of change of species i.

s: the number of species.

A generalized Lotka–Volterra model following Roberts (1974) and Tregonning and Roberts (1979—referred to by them as the multispecies quadratic model) is summarized for any number of species by:

$$\frac{dN_i}{dt} = N_i\left(r_i + \sum_{i=1}^{s}\alpha_{ij}N_j\right)$$ (5.5)

where r_i is positive for a producer (prey, competitor) and negative for a consumer (predator, pathogen) following the convention in Chapter 3. Therefore consumers exponentially decline in the absence of producers (Eqn 3.12). Producers show density-dependent regulation as illustrated by the one-species version of Eqn 5.5:

$$\frac{dN_1}{dt} = N_1(r_1 + \alpha_{11}N_1)$$

or

$$\frac{dN_1}{dt} = r_1N_1 + \alpha_{11}N_1^2$$ (5.6)

This equation is equivalent to the logistic equation $dN_1/dt = r_1N_1 - r_1N_1^2/K$ with α_{11} equal to $-r_1/K$. It is also apparent why the model is the (multispecies) quadratic model *sensu* Tregonning and Roberts (1979). In Chapter 3 it was shown that populations described by the logistic equation had a stable equilibrium of K. The equilibrium occurs at $dN/dt = 0$. Therefore for Eqn 5.6:

$$0 = rN_1^* + \alpha_{11}N_1^*N_1^*$$

where N_1^* is the equilibrium population size. Divide both sides by N_1^* and rearrange to give:

$$N_1^* = -r/\alpha_{11}$$

As a check recall that α_{11} in Eqn 5.6 equals $-r_1/K$, therefore $K = -r_1/\alpha_{11}$. As r must be positive for a single species, α needs to be negative to give a positive value of K or N_1^*. The sign of α is important and we will return to this later. (Note that there are two solutions to the equilibrium equation—the other is the trivial solution of $N^* = 0$. We lost that solution by dividing by N_1^*.)

With two species, Eqn 5.5 gives:

$$\frac{dN_1}{dt} = N_1r_1 + \alpha_{11}N_1N_1 + \alpha_{12}N_2N_1$$ (5.7)

$$\frac{dN_2}{dt} = N_2r_2 + \alpha_{21}N_1N_2 + \alpha_{22}N_2N_2$$ (5.8)

We can compare the parameters α_{11}, α_{12}, α_{21} and α_{22}, to the parameters in the competition equations in Eqn 5.1 and 5.2 and the predator–prey equations in Chapter 3. In comparison to the competition equations: $\alpha_{11} = -r_1/K_1$ (this is the result we noted for the one-species example in Eqn 5.6); $\alpha_{12} = -\beta_{12}r_1/K_1$, $\alpha_{21} = -\beta_{21}r_2/K_2$ and $\alpha_{22} = -r_2/K_2$. Generalizing for species i and j, $\alpha_{ii} = -r_i/K_i$ and, substituting for α_{ii}, $\alpha_{ij} = \beta_{ij} \alpha_{ii}$, i.e. the competition coefficient multiplied by the intraspecific interaction coefficient. Compared to the predator–prey equations (assume N_1 is prey and N_2 is predator): $\alpha_{11} = -r_1/K_1$; $\alpha_{12} = -\alpha$; $\alpha_{22} = 0$ and $\alpha_{21} = \beta$. Also r_2 will be negative and r_1 will be positive. Note that the interspecific interaction coefficients for competition are (– –) and for predation are (– +).

We can therefore begin to see how, with different values and signs of r and α_{ij}, Eqn 5.5 can provide a generalized description of Lotka–Volterra dynamics covering interactions such as interspecific competition and predation.

The community matrix is then derived by considering the community at equilibrium. If we take the two-species example:

$$0 = N_1 *r_1 + \alpha_{11}N_1 *N_1 * + \alpha_{12}N_2 *N_1 * \tag{5.9}$$

$$0 = N_2 *r_2 + \alpha_{21}N_1 *N_2 * + \alpha_{22}N_2 *N_2 * \tag{5.10}$$

We can evaluate the species densities at equilibrium:

$$-r_1 = \alpha_{11}N_1 * + \alpha_{12}N_2 *$$

$$-r_2 = \alpha_{21}N_1 * + \alpha_{22}N_2 *$$

In matrix form this is:

$$\begin{pmatrix} -r_1 \\ -r_2 \end{pmatrix} = \begin{pmatrix} \alpha_{11} & \alpha_{12} \\ \alpha_{21} & \alpha_{22} \end{pmatrix} \begin{pmatrix} N_1 * \\ N_2 * \end{pmatrix} \tag{5.11}$$

The values of N_1* and N_2* can be calculated using matrix algebra by finding the inverse of the matrix of coefficients and multiplying both sides to give:

$$N_1* = \left(\frac{-\alpha_{22}r_1 + \alpha_{12}r_2}{\alpha_{11}\alpha_{22} - \alpha_{21}\alpha_{12}} \right) \tag{5.12}$$

$$N_2* = \left(\frac{\alpha_{21}r_1 - \alpha_{11}r_2}{\alpha_{11}\alpha_{22} - \alpha_{21}\alpha_{12}} \right) \tag{5.13}$$

Matrix Eqn 5.11 can be generalized for any number of species as:

$$-\mathbf{r} = \mathbf{A}\,\mathbf{N}*$$

where $\mathbf{A}$ is the square matrix of interaction coefficients and $\mathbf{r}$ and $\mathbf{N}*$ are column vectors of intrinsic rates of change (r_i) and equilibrium densities (N_i*), respectively. Equilibrium values can then be found by matrix algebra as for Eqn 5.11 (equivalent to the solution of s simultaneous equations where A^{-1} is the inverse of A):

$$-r\,A^{-1} = N*$$

To determine the community matrix **M** we need to linearize the population growth equation (5.5) at equilibrium. This is achieved with a Taylor expansion or series. This is followed by evaluation of the partial derivatives at equilibrium to give the coefficients of the community matrix. The principles of Taylor expansion and partial differentiation are given in Box 5.1. Details of this method are given in Pielou (1977) with a further example in Chapter 7. Beginning with Eqn 5.5 repeated here:

$$\frac{dN_i}{dt} = N_i\left(r_i + \sum_{i=1}^{s} \alpha_{ij} N_j\right)$$

This is rewritten in general terms as:

$$\frac{dN_i}{dt} = F_i(N)$$

The Taylor expansion around the equilibrium is:

$$\frac{dN_i}{dt} = F_i(N*) + \sum_{j=1}^{k} n_j \left[\frac{\partial\left(\dfrac{dN_i}{dt}\right)}{\partial N_j}\right]$$

+ second and higher-order terms

where N_j is a small perturbation from equilibrium (see h in Box 5.1).

$F_i(N*)$ is 0 and second- and higher-order terms are ignored. The partial derivatives $\partial(dN_i/dt)/\partial N_j$ at equilibrium are conveniently:

$$\alpha_{ij} N_i*$$

i.e. the interaction coefficient multiplied by the equilibrium population density of the ith species. You could check this for the two-species example (Eqns 5.7 and 5.8). For example, to find $\partial(dN_1/dt)/\partial N_1$ at $N*$:

$$\frac{\partial\left(\dfrac{dN_1}{dt}\right)}{\partial N_1} = r_1 + 2\alpha_{11} N_1* + \alpha_{12} N_2*$$

Substitute for N_1* and N_2* (from Eqns 5.12 and 5.13 above) to give:

$$r_1 + \frac{2\alpha_{11}(-\alpha_{22}r_1 + \alpha_{12}r_2)}{\alpha_{11}\alpha_{22} - \alpha_{21}\alpha_{12}} + \frac{\alpha_{12}(\alpha_{21}r_1 - \alpha_{11}r_2)}{\alpha_{11}\alpha_{22} - \alpha_{21}\alpha_{12}}$$

This reduces to:

$$\frac{\alpha_{11}(-r_1\alpha_{22}+\alpha_{12}r_2)}{\alpha_{11}\alpha_{22}-\alpha_{21}\alpha_{12}}$$

which is $\alpha_{11}N_1{}^*$.

The full community matrix **M** for the two-species example is:

$$\begin{pmatrix} \alpha_{11}N_1{}^* & \alpha_{12}N_1{}^* \\ \alpha_{21}N_2{}^* & \alpha_{22}N_1{}^* \end{pmatrix}$$

The stability of the community is found by determining the eigenvalue(s) of the community matrix. This tells us about the growth of a perturbation (n_j) from equilibrium. If the sign of the largest eigenvalue of the community matrix is negative then the community is stable, i.e. the perturbations reduce in size back towards the equilibrium. A positive value indicates growth of the perturbation away from the equilibrium. Therefore we can see why the community matrix is sometimes referred to as the stability matrix. The magnitude of the dominant eigenvalue determines the return time of the community (Pimm & Lawton 1977) which measures the time taken for a perturbation to decay to 1/e of its initial value. (Note that dividing both sides of Eqn 5.5 by N_i will give the *per capita*, i.e. per individual, rates of population change. When this is done the general equation 5.5 is written as (see p.125):

Box 5.1 Partial differentiation and the Taylor series

Partial differentiation

We have used differentiation to find a gradient at a point on a curve. Now imagine that this curve is produced by cutting through a two-dimensional plane (Fig. 1). The gradient at the point (x_1, y_1) will depend on the direction in which the plane is cut. With partial differentiation we can imagine that the plane is cut at right angles along the (independent) horizontal x and y axes.

Finding dz/dx for constant y and dz/dy for constant x will give the gradients in the directions x and y respectively. These are referred to as partial derivatives and written with 'curly' d's:

$$\frac{\partial z}{\partial x} \text{ and } \frac{\partial z}{\partial y}$$

An ecological example would be where the abundance of a species is represented by z whilst x and y are the coordinates on the ground. At each point

Continued on page 122

Box 5.1 (*continued*)

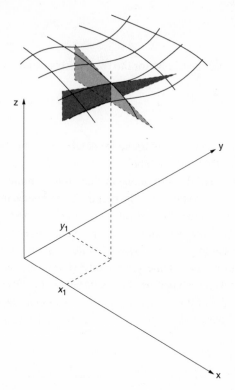

Fig. 1 Representation of gradients at x_1, y_1 achieved by cutting through plane at right angles.

x and y the rate of change in abundance in each direction is given by $\partial z/\partial x$ and $\partial z/\partial y$. There may be more than two explanatory variables, perhaps including time.

Consider the example:

$$z = 3x^2 + 4y \tag{1}$$

To find the partial derivative $\partial z/\partial y$ differentiate Eqn 1 holding x constant:

$$\frac{\partial z}{\partial y} = \frac{\mathrm{d}(3x^2)}{\mathrm{d}y} + \frac{\mathrm{d}(4y)}{\mathrm{d}y}$$

$\mathrm{d}(3x^2)/\mathrm{d}y = 0$ (because x is a constant) and $\mathrm{d}(4y)/\mathrm{d}y = 4$

Continued

Box 5.1 (*continued*)

Therefore $\dfrac{\partial z}{\partial y} = 4$

Similarly $\dfrac{\partial z}{\partial x} = 6x + 0 = 6x$

Taylor series

A series is defined in mathematics as the sum of a sequence of numbers. Such a sequence may arise arithmetically, e.g.

1, 3, 5, 7, ... each term is found by adding 2 to the previous term.

or geometrically

1, 2, 4, 8, 16 ... each term is found by multiplying the previous term by 2.

These sequences can be written in general terms and summed. For a geometric series the general version is given as:

$a + ar + ar^2 + ar^3 + \ldots$

where a is the first term (1 in the above example) and r is the common ratio (2 in the above example).

From an ecological perspective, series are important in representing geometric growth. They are analytically useful because various functions, such as e^x or $\sin(x)$ can be expressed as a series. To understand this, assume that e^x can be expressed as a geometric series:

$e^x = a + bx + cx^2 + dx^3 + \ldots$

To find a we substitute $x = 0$:

$e^0 = a$, i.e. $a = 1$.

To find the coefficients $b, c, d \ldots$ we need to differentiate both sides with respect to x. Conveniently $d(e^x)/dx = e^x$ and so after differentiating:

$e^x = 0 + b + 2cx + 3dx^2 + \ldots$

Again, substitute $x = 0$

$e^0 = b$, therefore b also equals 1.

Differentiate again:

$e^x = 2c + 6dx + \ldots$

Continued on page 124

Box 5.1 (*continued*)

and substitute $x = 0$

$$e^0 = 2c, c = 0.5$$

and so on.

This process of converting a function into a series is given by the Maclaurin series for a function $f(x)$:

$$f(x) = f(0) + x f'(0) + \frac{x^2 f''(0)}{2!} + \frac{x^3 f'''(0)}{3!} + \ldots$$

i.e. $f(x) = a + xb + x^2 c + x^3 d + \ldots$

where $f(0)$ means the function evaluated at $x = 0$ (e.g. e^0); $f'(0)$ means the derivative of the function evaluated at $x = 0$; $f''(0)$ means the second derivative evaluated at 0 and so on. 3! means 3 factorial which is $3 \times 2 \times 1$. To check the coefficients from the Maclaurin series against our values:

$$a = f(0) = e^0 = 1; \quad b = f'(0) = e^0 = 1$$

$$c = \frac{f''(0)}{2!} = 1/2$$

The more terms of the series we add up, the closer we get to the true value of e^x. Thus the series converges towards e^x. Consider $x = 1$, i.e. e^1 or e:

with one term e^1 is approximated as 1,

with two terms e^1 is approximated as $1 + 1 = 2$,

with three terms e^1 is approximated as $1 + 1 + 0.5 = 2.5$,

and with four terms e^1 is approximated as $1 + 1 + 0.5 + 1/6 = 2.667$. (The value of e to three decimal places is 2.718.)

The Maclaurin series is a special case of the Taylor series which expresses the function $f(x + h)$ as:

$$f(x + h) = f(x) + h f'(x) + \frac{h^2}{2!} f''(x) + \frac{h^3}{3!} f'''(x) + \ldots$$

where x and h are both variables.

In ecology the Taylor series is useful for providing a linear approximation to a function when h is small relative to x. For example, if x represents an equilibrium density of a population which is described by a non-linear function (e.g. a quadratic) then when h is small (a perturbation from the equilibrium)

Continued

Box 5.1 (*continued*)

the function near the equilibrium can be expressed according to the Taylor series as:

$$f(x + h) = f(x) + h f'(x) \tag{2}$$

i.e. ignoring h^2 and higher-order terms because h is relatively small. Equation 2 describes the linear tangent at equilibrium. In ecological communities we may be dealing with the abundance of a number of species, all of which have an equilibrium. Linearized dynamics at equilibrium are therefore represented by partial derivatives in the Taylor series—see text for details.

$$\frac{dN_i'}{dt} = r_i + \sum_{i=1}^{s} \alpha_{ij} N_j$$

where N_i' is per capita. Thus the *per capita* equivalent of Eqn 5.5 is linear.)

To conclude this section we link up the graphical interpretation of stability of the logistic equation in Section 3.3 (Chapter 3) with the analytical method of the community matrix, following May (1973a) and Pimm (1982). Recall that there are two equilibria with the logistic equation ($N^* = 0$ and $N^* = K$). To examine the stability of those equilibria in Chapter 3 (Section 3.3 on harvesting) we used a graphical method to examine perturbations (displacements) from equilibrium and asked whether those displacements would become larger with time. If the perturbations do become larger then the equilibrium is locally unstable.

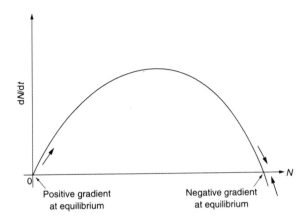

Fig. 5.4 Logistic curve showing unstable and stable equilibria. Note the gradient of the curves at $N^* = 0$ and $N^* = K$.

From the community matrix analysis we expect that the stable equilibrium of the logistic model is given by a negative slope of dN/dt with respect to N at equilibrium, i.e. the single 'eigenvalue' is negative. This is indeed the case: at $N^* = K$ the gradient is negative (Fig. 5.4, p.125).

Question 5.2

Summarize the assumptions of the community matrix.

5.2.3 Predictions from the community matrix

Several important results have emerged out of analytical and simulation studies of generalized Lotka–Volterra or similar community models. Gardner and Ashby (1970) asked whether large linearized systems (biological or otherwise) which were assembled at random would be stable. In so doing they were anticipating much of the current trend for analysing complexity; in their words they were concerned with 'an airport with 100 planes, slum areas with 10^4 persons or the human brain with 10^{10} neurons … [where] stability is of central importance'. They showed for small numbers of species/aeroplanes/neurons (n) that stability declines with connectance between components and as the number of components increases the system moves rapidly to a situation where there is a break-point (a step function) in which a small change in connectance will result in a switch from stability to complete instability (Fig. 5.5).

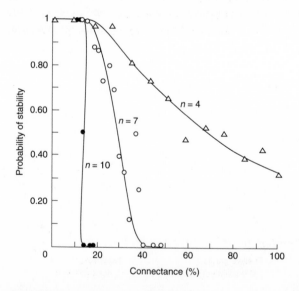

Fig. 5.5 Breakpoint/threshold of stability with connectance for a linear system with different numbers of components (n). From Gardner and Ashby (1970).

May (1972) discussed and generalized these results showing that increased numbers of species did not necessarily imply increased stability of the community and, indeed could produce the opposite result. Increased stability with complexity was believed by Elton (1958) whose conclusions were partly based on detailed case studies of invasions by 'pests' such as the giant snail *Achatina fulica* into Hawaii and the red deer *Cervus elaphus* into New Zealand, which contributed to dramatic declines in the endemic species of those islands. (See May (1984) and Pimm (1984) for a critique of Elton's position.) May demonstrated that a species which interacts widely with many other species (high connectance) should do so weakly (small α) and conversely those species which interact strongly with others should do so with a smaller number of species. He predicted that communities which are compartmentalized into blocks (effectively communities within communities, for example, guilds) may be stable whilst the whole may not be. Pimm (1984) provides a useful review of the meanings and relationship between stability and complexity.

Tregonning and Roberts (1979) explored these ideas further by examining the dynamics of a randomly constructed community in which the interaction coefficients, α_{ij}, were non-zero and chosen randomly and r was $+ 1$ or -1 (and therefore producers or consumers) at random with equal probability. They began with 50 species, ran the model and used two methods of species elimination: the species was either chosen at random or the species with the most negative equilibrium value was selected, i.e. they removed the most ecologically or physically unrealistic species, as all species needed to have a positive equilibrium value. This process was continued until all species had a positive equilibrium value. This was defined by Tregonning and Roberts as the **homeostatic system**—one which was ecologically feasible and at equilibrium. Under selective removal the mean number of species comprising a homeostatic system was 25 and the largest was 29. However, if elimination was random then the largest homeostatic system was 4 and the mean was 3.3.

Whilst there is no doubt that these predictions for community structure and dynamics are very exciting, it is easy to get carried away on the crest of a theoretical wave. In reality there have been few studies of these ideas in the field, partly owing to the difficulties of parameter estimation and partly because of the problems of communicating the theoretical ideas to practitioners in the field. In the next section we highlight a few examples where theory and field work have been combined.

5.2.4 *Estimations of community stability and structure in the field*

Seifert and Seifert (1976) have provided one of the few field tests of the community matrix using insects in Neotropical *Heliconia* flowers. These insects inhabited the water-filled bracts of two *Heliconia* species (*H. wagneriana* and *H. imbricata*) in

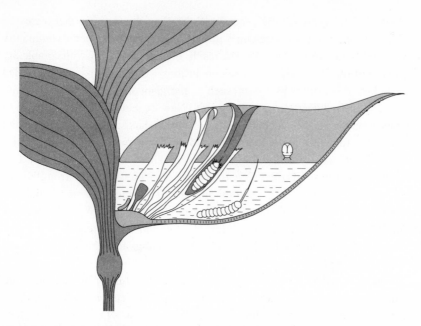

Fig. 5.6 Stylized view of *Heliconia wagneriana* showing the dissected bract with common inquiline insects. The species shown include *Gillisius* (located on the dissected bract just above the water), *Quichuana* (located at the base of the flower below *Gillisius*), *Copestylum* (located inside the flower) and *Beebeomyia* (located at the base of the seed). From Seifert and Seifert (1976).

rain forests in Costa Rica (Fig. 5.6). The insect species included Chrysomelid beetles and Syrphid flies, all of which were potential competitors.

Seifert and Seifert combined experimental manipulations with a multiple regression method which allowed them to estimate the magnitude and signs of the interaction coefficients from a generalized Lotka–Volterra model. This meant that statistical significance could be attached to each of the coefficients and therefore they were able to achieve the desirable simplest realistic model outlined in Chapter 1. The experiment involved emergent buds of *Heliconia* being enclosed in plastic bags to restrict immigration and oviposition. After a certain amount of growth, water was added and varying amounts of four species of insect were introduced. Following this the *per capita* change in numbers with time was determined using the linearized version of Eqn 5.5, calculated as the change from initial density divided by the number of days over which the change took place. The initial densities of each species were used as the explanatory variables to calculate the (partial) regression coefficients of the *per capita* rates of change against all species, i.e. this gave r and α_{ij}. (Note that the rates of change were not estimated from equilibrium as assumed by the community matrix.) A negative value of the regression coefficient indicated competition whilst a positive value indicated mutualism (the possibility of predation was ignored given the choice of insects). From Table 5.2 we see that

Table 5.2 Interaction matrix for *Heliconia wagneriana* (see Fig. 5.6 for details of species). Non-significant elements are set to zero (data from Seifert & Seifert 1976).

	Quichuana	Gillisius	Copestylum	Beebeomyia
Quichuana	0.001	0	–0.018	0.027
Gillisius	0	–0.003	0	0
Copestylum	0	0	–0.005	0
Beebeomyia	0	–0.005	0	–0.033

nine of the interactions were not significant (and therefore set to 0). Of the significant interspecific ones, two were negative (competitors) and one positive (mutualism).

The equilibrium densities estimated from the model by $N^*_i = A^{-1}r_i$ are shown in Table 5.3 compared with those observed in the field. The fact that there are two negative (unrealistic) densities for *H. wagneriana* suggests either that the observed mean densities are not equilibrium densities, or are results of processes not dependent on species interactions, or that the model is inappropriate. Seifert and Seifert were then able to determine the eigenvalues of the community matrix composed of $\alpha_{ij}N^*_i$ and therefore examine the stability of the community. The four values of the eigenvalues were: –0.221, 0.052, –0.042 and –0.239, i.e one positive eigenvalue indicating an unstable community. Therefore *H. wagneriana* communities were judged by Seifert and Seifert not to be stable. Instead they concluded that migration, oviposition and local extinction processes may be important in structuring these communities. In other words it is probably not correct to model these communities in isolation. Migration and local extinction processes are considered in Chapter 6.

Wilson and Roxburgh (1992) provided examples of the application of the community matrix to plant species mixtures. They predicted that initially unstable six-species mixtures will, by selective deletion (following Tregonning & Roberts 1979), drop down to stable four-species mixtures. A study of the persistence of chironomid communities in the River Danube demonstrated differences in return times of perturbed communities at different sites (Schmid 1992). An analysis of local and global stability in six small mammal communities (Hallett 1991) showed that all the community matrices were locally and globally stable. The latter was

Table 5.3 Equilibrium densities predicted from the model and mean densities observed in the field (data from Seifert & Seifert 1976).

densities	Mean densities in unmanipulated examples	Estimated species equilibrium
Quichauna	51.00	–112
Gillisius	7.56	–23.2
Copestylum	8.78	4.09
Beebeomyia	6.67	10.62

due to a reduction in connectance with increasing number of species (which would be predicted by the model of Gardner & Ashby, (1970))

The examples in this section show that it is possible to parameterize community matrix models using field data (with or without manipulations) and make testable predictions about stability, structure and return times after perturbation. Furthermore, comparative work such as that of Hallett (1991) allows these predictions to be related to species richness and connectance. However, we need to be cautious as analysis of the community matrix is in the neighbourhood of equilibrium. For many applications we are likely to be interested in communities away from equilibrium (assuming they have one) and where non-equilibrium processes such as physical disturbance or pollution events may be important. Local extinction and colonization processes may also mean that equilibrium has to be judged at larger spatial scales, perhaps in the context of metapopulations rather than local populations (Chapter 6).

5.3 Models of succession

Succession is the directional change in plant and animal species over time in a particular area. Mathematical models of this phenomenon have represented it as a **Markov chain** (Horn 1975, 1981). This involves determining the probability that a given plant (or other species or suite of species) will be replaced in a specified time by another individual(s) of the same or different species. It is assumed that these replacement probabilities *do not* change with time. Starting with a particular distribution of species, the relative abundances of species are multiplied by the transition probabilities to generate new relative abundances. This is then repeated a certain number of times. For example, Horn (1975, 1981) gave the values for 50-year tree-by-tree replacement between four species (Table 5.4a).

The model can be represented in matrix form:

$$\begin{pmatrix} GB \\ B \\ RM \\ M \end{pmatrix} = (\text{transition probabilities}) \begin{pmatrix} GB \\ B \\ RM \\ M \end{pmatrix}$$

$\mathbf{v}_{t+1}$ transition probability matrix $\mathbf{v}_t$

These models predict a **stationary end-point**, i.e. that there will be a fixed ratio of *GB* to *B* to *RM* to *B* (see Question 5.3). This is analagous to the result of a stable age structure in a structured population (Chapter 4). Iterations over different periods of time and the end-point of the Horn example are given in Table 5.4(b). The predicted end-point compares favourably with the observed composition in old growth forest.

The study of succession using Markovian processes has been revitalized

Table 5.4a Fifty-year tree-by-tree transition matrix for grey birch, blackgum, red maple and beech.

	50 years hence			
Now	Grey birch	Blackgum	Red maple	Beech
Grey birch	5	36	50	9
Blackgum	1	57	25	17
Red maple	0	14	55	31
Beech	0	1	3	96

Table 5.4b Predicted composition of a succession. Table 5.4 from Horn (1981).

	Age of forest (years)						Very old forest
	0	50	100	150	200	... ∞	
Grey birch	100	5	1	0	0	0	0
Blackgum	0	36	29	23	18	5	3
Red maple	0	50	39	30	24	9	4
Beech	0	9	31	47	58	86	93

by the increasing availability of remote sensed data, Geographic Information Systems and rapid processing speed on personal computers allowing simulations of large-scale and long-term ecological processes. The next two examples illustrate landscape-scale studies of succession leading into the overview of spatial dynamics in Chapter 6.

Frelich *et al.* (1993) examined the long-term effects of tree-by-tree replacement processes on spatial patterns in a forest. Transition probabilities were found to be dependent on the species composition of a local neighbourhood (10 m radius). Frelich *et al.* determined whether sugar maple (*Acer saccharum*) or hemlock (*Tsuga canadensis*) trees had a positive, negative or no effect on nearby establishment of understory trees of other species. They did this by examining overstory composition in 38 randomly located plots and relating it to understory composition. It was found that intraspecific associations were positive and interspecific associations negative. In their model the plots were square, 16 ha in size, with each simulated 'tree' mapped. The density of trees was comparable to that observed in the field (400 canopy trees per ha). The model was run for 3000 years (the length of time the sugar maple–hemlock mixture had been in existence) with 10-year time steps. Three events occurred at each time step as follows.

1 The probability of death was determined for each tree.

2 A new canopy tree was recruited to replace each dead tree. The new tree was placed on the map of individuals within a 4 m radius of the dying tree. The species of new recruit depended on the species composition of the local neighbourhood

..ld the observed interactions within and between species.

3 The age of all trees was increased by 10 years.

Simulations were undertaken demonstrating that the patches of sugar maple and hemlock observed in the field could have developed from an initial random mix and without an underlying basis of topography, soil or disturbance. After 3000 years, small single-species groups of trees that initially occurred by chance had been magnified to become major patches (Fig. 5.7a). This was in agreement with field studies which had failed to detect reasons for any pattern.

Frelich *et al.* also simulated the effects of random invasion of sugar maple by hemlock, assuming a 1% probability of replacement of sugar maple by hemlock, regardless of local species composition, i.e. they changed the interspecific competition rules but maintained the intraspecific rules as above (Fig. 5.7b—their experiment 2). The invasion simulation was then reversed (Fig. 5.7c), starting with

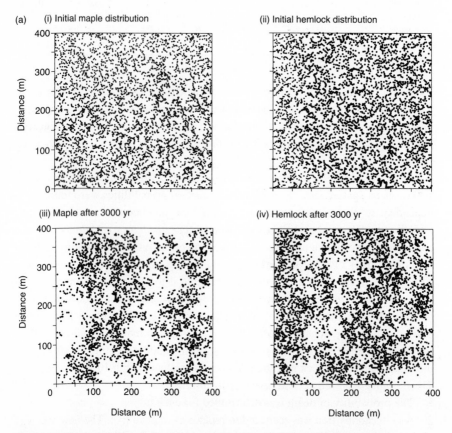

Fig. 5.7 The hemlock and maple distribution; (a) shows in part (i) the initial sugar maple distribution, (ii) the initial hemlock distribution, (iii) sugar maple distribution after 3000 yr and (iv) hemlock distribution after 3000 yr.

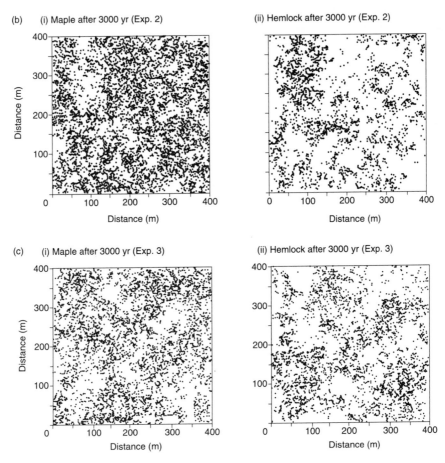

Fig. 5.7 (*continued*) (b) Shows sugar maple and hemlock distribution for experiment number 2 and (c) shows sugar maple and hemlock distribution after experiment number 3.

three clusters of hemlock, each 20 m in diameter and with the same rules as (b). This simulated initial aggregation of hemlock in local patches (favourable microhabitats) followed by climate change allowing hemlock to grow anywhere (Fig. 5.7c—their experiment 3).

Other studies have looked at successional transitions between woodland and other types of vegetation. For example, Callaway and Davis (1993) used aerial photographs to measure transition rates between grassland, coastal sage scrub, chaparral and oak woodland and their relationship to burning and grazing in Gaviota State Park in central coastal California between 1947 and 1989. The percentages of vegetation (community) types in 1947 and 1989 are given in Table 5.5 based on plots marked on aerial photographs. These plots corresponded to 0.25 ha on the ground.

Table 5.5 The percentage of vegetation type from aerial photographs in 1947 and 1989 in central coastal California.

Year	Vegetation type (%)			
	Grassland	Coastal sage	Chaparral	Oak woodland
1947	21.5	26.4	28	24.1
1989	23.3	25.9	24	26.8

Although the overall percentage cover was very similar there was considerable flux between the years within plots. Transition between vegetation type occurred in 71 out of 220 plots (32%). The transition probabilities were determined using these data (Fig. 5.8).

Once the transition probabilities were known the current state (e.g. oak woodland) could be multiplied by the four transitions (including no change) in the 42-year period. This was then repeated in a Markov chain over time to predict the change in vegetation under particular environmental conditions. The predictions

Question 5.3

Given an initial abundance of tree species of 10 grey birch, 10 blackgum, 10 red maple and 10 beech, use the transition probabilities in Table 5.4(a) to predict the relative abundance of these species after one transition (50 years).

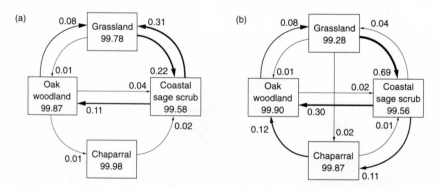

Fig. 5.8 (a) Annual transition rates among plant communities in burned plots ($n = 53$) as determined from changes in vegetation between 1947 and 1989 shown on aerial photographs. (b) Annual transition rates among plant communities in unburned plots ($n = 78$) as determined from changes in vegetation between 1947 and 1989 shown on aerial photographs. The numbers in the boxes estimate the probability, as a percentage, that a given community will remain the same; the numbers on the arrows estimate the probability that a community will change in the indicated direction (thickness of lines is proportional to the probability of that change).

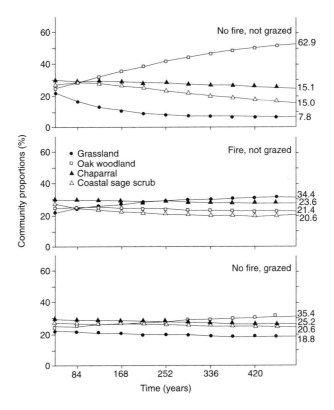

Fig. 5.9 Markov chain model predictions of future change in proportions of plant communities. Final community proportions at stability (defined as <0.1% change over 42 yr) are presented at the right, opposite each curve.

for three combinations of burning and grazing are shown in Fig. 5.9.

The greatest changes in vegetation were predicted to occur under conditions of no grazing and no fire.

5.4 Conclusions

The community models presented in this chapter pose important challenges to the theoretician and practitioner. On the one hand analysis of community models requires mathematical techniques which are complex and assumptions (such as linearized dynamics in the neighbourhood of the equilibrium) which are at best problematic. On the other hand critical field studies require detailed knowledge of life-histories, dynamics of species and experimentation. The work of Seifert and Seifert shows that combinations of these skills are possible. The predictions of community models are so exciting that ecologists must pursue their study in the field. For the successional models, although time-scales are prohibitive, predictions of models can be tested spatially. In the next chapter we develop the spatial dimension of modelling.

CHAPTER 6

Spatial models and thresholds

So far all the models in Chapters 2–5 have been considered at one spatial scale. Most of these, with the exception of the Markovian succession model, have concentrated on small-scale dynamics. Implicit in such models are the assumptions that all individuals are in close proximity and mixing freely, that spatial heterogeneity is not important and that populations are closed, i.e. that there is no immigration into or emigration from these populations. Clearly these assumptions are unrealistic for many populations. Models incorporating a spatial element are becoming increasingly popular—one reason for this is the development of computer software such as Geographic Information Systems and the increased memory and processing speed on personal computers required for data handling (as illustrated for successional processes in Chapter 5). A second reason is that incorporation of spatial processes has revealed and continues to reveal results on stability and dynamics that were unpredicted from local population models. These spatially explicit models have important consequences for species persistence and the effects of large-scale anthropogenic change such as regional and global climate change. The simple models described in this chapter provide an insight into the effects of the spatial dimension(s).

6.1 Spatial dynamics of host–parasitoid systems

6.1.1 Introduction to host–parasitoid systems

It has been estimated that more than 10% of metazoan animals are parasitoids (Hassell & Godfray 1992). Parasitoids are wasps or flies which lay their eggs on or inside a host larva, such as a moth or a fly. The hosts are often herbivorous insects, although there are also parasitoids of parasitoids called hyperparasitoids. The parasitoid larva then feeds on the host larva and eventually kills the host. Given the number of parasitoid species it is important to understand the type of dynamics which may be produced by host–parasitoid interactions.

We will explore the earliest model of host–parasitoid dynamics devised in the 1930s and see how this has been modified in the 1990s to include spatial dynamics.

6.1.2 Nicholson–Bailey model

The earliest model of host–parasitoid interactions was constructed by Nicholson

& Bailey (1935) who made a series of assumptions and calculations, several of which still feature in models developed in the 1990s. Their model was important as it made the case for time delays described by discrete time equations rather than the continuous time Lotka–Volterra equations which were derived a few years earlier (Chapter 3). The main assumptions of the Nicholson–Bailey model are given below.

1 Either zero or one parasitoid is produced per host (even if more than one egg is laid on or in a host).

2 Each female parasitoid searches a fixed area a, finding all the hosts. Therefore the probability of a host being attacked is a/A where A is the total study area and the probability of not being parasitized is $1 - a/A$. If P is the density of parasitoids then there are AP female parasitoids. If parasitoids search independently (and at random) then the probability of a host not being attacked by any parasitoid is given by the Poisson distributions (Box 6.1). If a/A is replaced by α (defined as the proportion of total hosts encountered by one parasitoid per unit time) then the probability of escaping attack is $e^{-\alpha P}$, i.e. the first term of the Poisson distribution. The probability of being attacked at least once is then $1 - $ (probability of not being attacked), i.e. $1 - e^{(-\alpha P)}$.

3 The finite rate of increase of hosts (λ_H, in the absence of parasitoids) and parasitoids (λ_p) and the density of hosts (H) are known.

Using the above reasoning produces a set of two equations, one for the dynamics of the host (H) and one for the dynamics of the parasitoid (P):

$$P_{t+1} = \lambda_p H_t \left(1 - e^{-\alpha P_t}\right) \tag{6.1}$$

$$H_{t+1} = \lambda_H H_t \, e^{-\alpha P_t} \tag{6.2}$$

Box 6.1 Examples of probability density functions

Discrete probability density functions (pdfs)

A random variable, X, is referred to as discrete if X can only take certain countable values. The probability density function (pdf) of X is then a set of mathematical statements which tell us the probability that X will take each of those values.

For example, if X can take any of the values 1, 2, 3, 4, 5 or 6 with equal probability (e.g. when rolling a dice), then X comes from a uniform distribution with a pdf as follows:

$$P(X = x) = \frac{1}{6} \text{ where } x = 1, 2, 3, 4, 5 \text{ or } 6.$$

Continued on page 138

Box 6.1 (*continued*)

It is also a rule that when all the values of a pdf are summed, they will come to 1; i.e. X must take one of the defined values.

A histogram can provide a graphical illustration of this pdf. In this simplest case of a uniform distribution, the pdf will be as in Figs 2.4 and 2.5.

Continuous pdfs

Continuous variables may also be defined in terms of their pdf. However, for a continuous variable (Y) it is not possible to speak of the probability of Y taking a specific value (e.g. the probability that $Y = 5.62$). This is defined as being so small that it is effectively zero. Instead, we must talk of the probability of Y lying within certain limits.

If all the thistles in a (very large) field site are defined as our population (i.e. these represent the full set of thistles that we are interested in), and the heights of all of these thistles have been measured, then when plotted on a graph the distribution may look in Fig. 1.

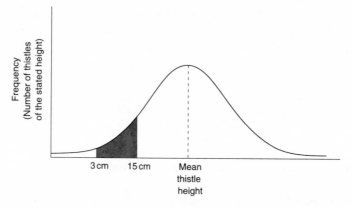

Fig. 1 The probability density function of thistle heights.

This is equivalent to the probability density function of thistle heights, for all thistles at the field site lie somewhere on this graph. We may then be interested in the proportion of thistles that lie between 3 and 15 cm (Fig. 1). If the graph is scaled so that the area under the curve is equal to 1, then the shaded area will equal the probability that any one thistle chosen at random from the population will lie between 3 and 15 cm.

If we have a mathematical expression which describes the curve, then this probability can be found directly by integration between the two limits of interest.

Continued

Box 6.1 (*continued*)

The Poisson distribution

A process by which events occur at random in space or time is known as a Poisson process. The distribution of those events (i.e. the number of events occurring per unit of time or space) is described by the Poisson distribution.

The Poisson distribution is an example of a discrete pdf as it is concerned with *counts* of events. A Poisson process is recognized by its properties of homogeneity and independence. By homogeneity, we mean that the probability of an event occurring per unit time or space remains constant—there is no tendency for the process to speed up or slow down in time or space. Independence refers to the fact that each event must be quite independent of any other.

For example, the emission of radioactive particles may follow a Poisson process, as long as the emission of one particle does not make it more or less likely that another particle will be emitted. The distribution of daisies on a lawn, however, is less likely to follow a Poisson distribution because daisies can reproduce asexually and seed may be distributed locally so that the presence of one daisy increases the probability that another daisy will appear in the close vicinity.

The pdf of the Poisson distribution is defined as $P(X = x) = e^{-\mu}\mu^{x}/x!$ where $x = 0, 1, 2$ etc. and $\mu =$ the mean number of events happening per unit time or space.

If the events in question are rare, μ will be low (e.g. 0.8), and the resulting pdf will take the form shown in Fig. 2.

Frequently, with $\mu = 0.8$, no events occur at all and it is very rare for more than four events to occur in one unit of time or space. The pdf has very pronounced right skew (Fig. 2) (the direction of skew always refers to the position of the tail).

If events occur rather more frequently (e.g. $\mu = 2$), the shape of the pdf will become less skewed (Fig. 3).

Finally, when the mean becomes quite high (e.g. $\mu = 6$) the distribution becomes much more symmetrical (Fig. 4).

The negative binomial distribution

For many biological processes, the assumptions required for the Poisson distribution to be a reasonable description of the data do not hold. For example, if we wish to describe the number of helminth parasites present per human in a village population, there may be various reasons for suspecting that the conditions of homogeneity and independence are broken. Some members of

Continued on page 140

Box 6.1 (*continued*)

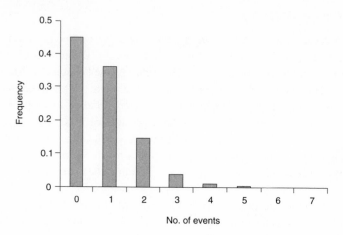

Fig. 2 The frequency and number of events where the pdf has a very pronounced right skew given by $\mu = 0.8$.

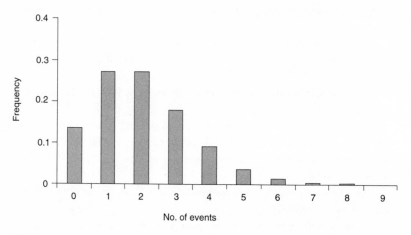

Fig. 3 The frequency and number of events where the pdf is less skewed given by $\mu = 2$.

the village may be living in conditions that expose them more frequently to the infectious stage of the parasite, or some may have a natural immunity. It may also be the case that once established in the human gut, the parasites reproduce. These sorts of processes will lead to some humans carrying heavier parasite loads than expected, and more people being parasite-free than expected. Such distributions are referred to as aggregated as, in this example, the parasitic

Continued

Box 6.1 (*continued*)

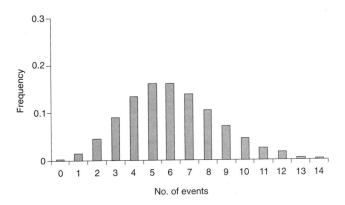

Fig. 4 The frequency and number of events where the distribution becomes symmetrical ($\mu = 6$).

worms are clumped together in their distribution across humans. A probability distribution that is frequently employed to describe such patterns is the negative binomial distribution, which has its pdf defined as follows:

$$P(X = x) = \left(1 + \frac{\mu}{k}\right)^{-k} \frac{(k + x - 1)!}{x!(k - 1)!} \left(\frac{\mu}{\mu + k}\right)^{x}$$

where μ is the mean number of parasites per human (just as it would be defined in the Poisson distribution). This distribution has an extra parameter, k. The reason that this distribution is so popular is that by altering the value of k, the degree of clumping can be altered. As k becomes infinitely large, the distribution converges on the Poisson distribution. As k becomes smaller, the distribution becomes more clumped; that is, the same number of parasites become distributed over fewer humans—more people being in the parasite-free class. For further details on how to chose the value of k, that best fits your data, see Elliott (1977).

Thus, the number of parasitoids at time $t + 1$ (P_{t+1}) is equal to the number of hosts at time t (H_t) multiplied by the fraction of hosts which are attacked ($1 - e^{-\alpha P t}$) multiplied by the finite rate of increase of the parasitoids (λ_p), whilst the number of hosts at time $t + 1$ (H_{t+1}) is equal to the number of hosts at time t (H_t) multiplied by the fraction of hosts which are not attacked ($e^{-\alpha P t}$) multiplied by the finite rate of increase of the host (λ_H).

Without host (prey) density dependence this system has an unstable equilibrium and produces cycles which can very easily become divergent and lead to the local extinction of the parasitoid. This has been tested in laboratory combinations of host and parasitoid and compared with predictions from the Nicholson–Bailey model (Fig. 6.1a).

One option to introduce host density dependence into the model is to multiply the finite rate of increase of the host (λ_H) by a linear term $1-H_t/K$, perhaps representing intraspecific competition amongst hosts. The linear density dependence term was discussed in Chapter 2 and in Chapter 3 where the Lotka–Volterra predator–prey system was stabilized in a similar way.

Beddington *et al.* (1975) examined the effect of including host density-dependent regulation using $H_{t+1} = H_t \, e^{[r(1-H_t/K)-\alpha P_t]}$ to replace Eqn 6.2, i.e. multiplying the intrinsic rate of increase r by a linear density dependence term. Stability was indeed much more likely and new types of dynamics were produced such as 5- and 20-point cycles. This was compared to the prey equation in the absence of predation which followed the standard period doubling route to chaos (Chapter 2). The production of cycles from Eqn 6.1 and modified Eqn 6.2 is expected from the discussion of delayed density dependence in Chapter 3 where the second-order Ricker equation gave cycles of period 6–7 years for the larch bud-moth. Manipulation of the parameters in these equations can produce cycles with different periods. Murdoch and Reeve (1987) discussed the role of direct and delayed density dependence in host–parasitoid models.

Other possibilities exist for stabilizing the host–parasitoid system, for example, the two assumptions of fixed search area and random searching by parasitoids have been shown to be unreasonable and to affect the stability of the interaction (Hassell & Godfray 1992). The effect of search area on stability was demonstrated by assuming competition between the parasitoids (Hassell & Varley 1969, Hassell & May 1973) which produces cycles or a stable equilibrium dependent on the intensity of interference (Fig. 6.1b). The effect of an aggregated distribution of parasitoids on host–parasitoid stability has been modelled using the negative binomial (May 1978, Box 6.1). Increased parasitoid aggregation is stabilizing because it increases the risk of mortality for hosts in areas of high host density.

The latter is an example of positive **spatial density dependence** where, at one point in time, densities of hosts are spatially variable. The density dependence discussed in previous chapters was temporal density dependence where densities fluctuated between years or other time periods. Field populations may be expected to show both temporal and spatial density dependence. The possibility of spatial density dependence provides us with another reason for producing spatially explicit models. However, it is not always the case that parasitoids show aggregation in response to local host density. The degree of parasitoid aggregation was determined from field data relating to the successful biological control of California red scale insects by their parasitoids (Reeve & Murdoch 1985). In this case no evidence was found for parasitoid aggregation at any spatial scale, despite the fact that it had

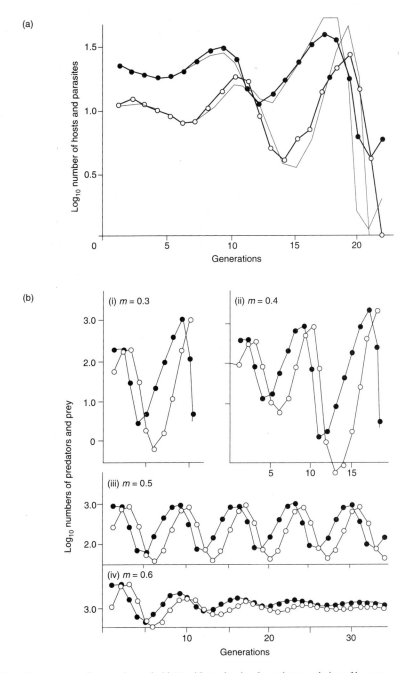

Fig. 6.1 Dynamics of host and parasitoid (a) without density-dependent regulation of host or parasitoid. Observed fluctuations from an interaction between the greenhouse whitefly, *Trialeurodes vaporariorum* (●), and its parasitoid wasp, *Encarsia formosa* (○). Thin lines show results of the Nicholson–Bailey model (after Burnett 1958). (b) With increasing levels of competition between parasitoids. Predator (○) and prey (●) oscillations from modified Nicholson–Bailey model showing the progressive stability as the interference constant (*m*) increases from 0.3 to 0.6 (from Hassell & Varley 1969). Figure reprinted from Hassell 1976.

been believed to stabilize the interaction. This was important because the ideal end-point in biological control is that both the host and the control agent remain at low equilibrium densities, as illustrated by the successful control of the prickly pear, *Opuntia stricta* by the moth *Cactoblastis cactorum* in Australia (discussed in Krebs 1994). Reeve and Murdoch suggested other reasons for the success of the red-scale system, including the fact that hosts had refugia which allowed certain of them to escape parasitoid attack and therefore prevent local extinction of the host. This introduces sparial density dependence without the need for parasitoid aggregations. The importance of prey refugia in stabilizing predator–prey dynamics was demonstrated by the classic experiments of Huffaker (1958) with oranges, herbivorous mites (prey) and predatory mites. Greater spatial heterogeneity led to increased persistence of predator and prey, by balancing migration rates of predator and prey, again emphasizing the importance of spatially explicit population models.

In the next section the role of spatial scale is explored with the basic Nicholson–Bailey model and the consequences for dynamics and stability are examined. Before continuing it is worth reflecting that Nicholson and Bailey recognized many of the possible developments of their model, leading to persistence. This included regulation of the prey (host):

> When the density of a species becomes very great as a result of increasing oscillation the retarding influence of such factors as scarcity of food or of suitable places to live is bound to be felt. Clearly these factors will prevent unlimited increase in density so … that the oscillation is perpetually maintained at a large constant amplitude in a constant environment.

and the anticipation of a spatial element:

> A probable ultimate effect of increasing oscillation is the breaking up of the species-population into numerous small widely separated groups which wax and wane and then disappear, to be replaced by new groups in previously unoccupied situations.

Question 6.1

Draw a flow diagram (similar to that of Fig. 2.8) to calculate the changing values of P and H using Eqns 6.1 and 6.2 (Nicholson–Bailey model).

6.1.3 Spatial dynamics with the Nicholson–Bailey model

Comins *et al.* (1992) used the Nicholson–Bailey model (Eqns 6.1 and 6.2) to explore the possible spatial dynamics of a host–parasitoid system. They assumed that the host and parasitoid were distributed amongst a grid of square cells or patches of width n. This spatial model framework, known generally as **cellular automata**, has been widely used in both plant and animal studies (e.g. Crawley & May 1987, Silvertown *et al.* 1992, Colasanti & Grime 1993, Perry & Gonzalez-Andajur 1993; see Wolfram 1984 for a mathematical overview). Comins *et al.* had two phases of dynamics: reproduction/parasitism and dispersal. The former was modelled using the Nicholson–Bailey model. The latter had the following simple rules.

1   A fraction of the hosts and parasitoids leave the patch (grid cell) and the remainder stay to reproduce in their patch.

2   The fraction dispersing are equally divided between the eight neighbouring patches. There is only one movement per generation. The model of Comins *et al.* specifically excluded longer range dispersal.

3   There are reflective boundary conditions in which dispersing individuals are prevented from crossing the boundary and remain in the edge patch. Thus there is an explicit edge effect in this model, in contrast to some other cellular automata models.

In small arenas of less than 10 cells × 10 cells extinction of host and parasitoid occurred within a few hundred generations of the simulation (in agreement with Fig. 6.1(a) showing the basic instability of the model without dispersal). However, when the arena size was increased to between 15 and 30 patches three general types of spatial dynamics were found which Comins *et al.* described as spirals, spatial chaos and crystal lattices (Fig. 6.2a). The key feature of the three dynamic types is that they permit long-term persistence of the host and parasitoid within a relatively narrow range of population densities (Fig. 6.2b) as did the incorporation of host density dependence or aggregation of parasitoids. Similar results were also found with the oscillatory unstable discrete version of the Lotka–Volterra model (Fig. 6.2a). Other workers have explored the role of spatial processes of predators and prey as a contribution to the stability of their dynamics. For example, McCauley *et al.* (1993) used an individual-based model to determine the relative importance of predator and prey mobility on stability.

In conclusion, a spatially explicit model can produce long-term population persistence in contrast to an unstable local population model. This work is still in its early stages, for example, we do not know the importance of the dispersal assumption (2) above, although labour intensive mark–release–recapture data collected in the field are beginning to shed light on the short- and long-distance movement of host and parasitoids. Jones *et al.* (1996), in a study of the movements of a tephritid fly (a thistle seed head feeder) and its parasitoids, showed frequent movements across a patch of thistles of about 50 m by 50 m. The parasitoids moved further than the hosts within the patch. Longer distance dispersal has been demonstrated by Dempster *et al.* (1995) using rubidium and other chloride salts in

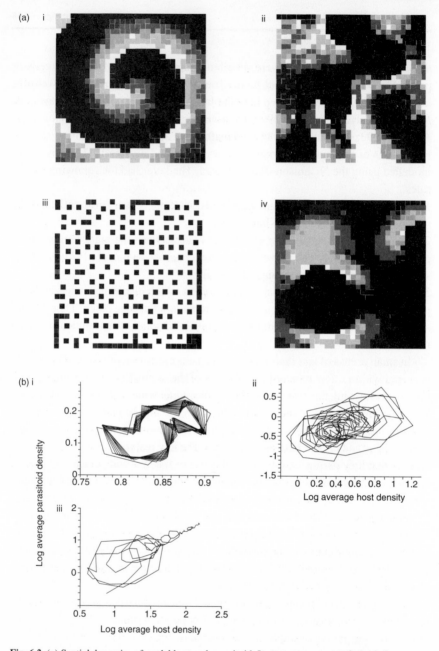

Fig. 6.2 (a) Spatial dynamics of model host and parasitoid. Instantaneous maps of population density for simulations of the dispersal model with Nicholson–Bailey local dynamics, with $\lambda = 2$ and arena width of 30 patches. Different levels of shading represent different densities of hosts and parasitoids. Black squares represent empty patches; dark shades becoming paler represent patches with increasing host densities; light shades represent patches with hosts increasing parasitoid densities. i, Spirals; ii, spatial chaos; iii, crystalline structures; case iv is a similar map obtained with Lotka–Volterra local dynamics. It exhibits highly variable spirals. (The maps are single frames from simulations over many generations.) (b) Phase plane showing changes of host and parasitoid over time with the same parameters as in (a i, ii, iii). From Comins *et al.* (1992)

plants to mark herbivores and parasitoids. This work demonstrated that distances of up to 800 m are not a barrier to colonization.

In the next section we develop the theme of spatial models, focusing on analytical techniques rather than on the results of cellular automata simulations.

6.2 Metapopulation models

6.2.1 Introduction to the metapopulation concept

A **metapopulation** is defined as a set of local populations linked by dispersal. This could be described and modelled by cellular automata but we will focus on results arising from simple analytical considerations. In the original model of Levins (1969, 1970) it was assumed that all local populations were of equal size and that a local population could either become extinct or reach carrying capacity following colonization instantaneously. Therefore only two states of local population were envisaged: full (carrying capacity) or empty (extinct).

In reality the definition of a local population (and therefore a metapopulation) is very difficult. Hanski and Gilpin (1991) define a local population as a 'set of individuals [of the same species] which all interact with each other with a high probability'. But how high is that probability? Also 'local' may be different for different interactions or the same interaction in different habitats; for example, two plants may show intraspecific competition over a scale of a few centimetres but be reproductively linked by pollination over hundreds of metres. It is also very difficult to say over what distance colonization of new areas (and therefore the 'birth' of new local populations) may occur. Typically the frequency of movements of propagules such as seeds over short distances are known, but longer distance movement (as noted in the previous section) is poorly known because it is often a rare event— excluding species which show seasonal and predictable long-distance migration.

Even when local populations can be identified, the pure Levins model of local populations with equal carrying capacity is unusual. More realistically, it is possible to envisage a spectrum of possibilities from mainland–island or core–satellite to pure Levins populations (Fig. 6.3). These and other possibilities have been discussed by Harrison (1991, 1994) who considered the rarity of true Levins metapopulations in the field and by Hanski and Gyllenberg (1993) who showed how to model both mainland–island and pure Levins with related equations.

Various processes will promote something close to a metapopulation structure in the field or at least create conditions under which local extinction and colonization are integral features of the population dynamics:

1 gap creation or other disturbance within non-recruitment habitat, including forests, rocky shores and grasslands;

2 a mosaic of successional habitats where, e.g., an annual plant must move from one transient early successional habitat to another;

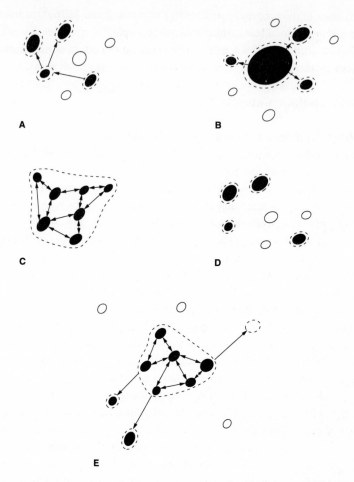

Fig. 6.3 Various types of spatial distribution of populations. Closed circles represent habitat patches; filled = occupied; unfilled = vacant. Dashed lines indicate the boundaries of 'populations'. Arrows indicate migration (colonization). (A) Levins metapopulation; (B) core–satellite (Boorman–Levitt, 1973) metapopulation; (C) 'patchy population'; (D) non-equilibrium metapopulation (differs from A in that there is no recolonization); (C) an intermediate case that combines (B) and (C). (From Harrison 1991.)

3 increasing fragmentation of habitats, and
4 sedentary and localized resources such as plants, dung or decaying logs, all of which may be attacked by various insects, fungi and other taxa.

6.2.2 The metapopulation model of Levins
Despite the problems of finding Levins metapopulations in the field it is instructive to consider how the simplest Levins metapopulation might behave before we go

on to related and more complicated models. Levins (1969, 1970) was interested in the number of islands or island-like habitats occupied by a species. Later Levins & Culver (1971) modified the model to investigate the effect of competition on the migration and extinction rates. Levins began by taking the number of local populations (N), the total number of sites (T), an extinction rate (e) and a migration rate (m'). The rate of change of N with time (t) could then be expressed by a differential equation as:

$$\frac{dN}{dt} = m'N(T - N) - eN$$

This equation was simplified by letting $p = N/T$ so that p represents the fraction of habitat patches occupied by a species and $m = m'T$. The rate of change in the fraction of habitat patches occupied by a species, dp/dt, i.e. the rate of change in the proportion of local populations (p, or strictly p_t) at a given time, was then described by:

$$\frac{dp}{dt} = mp(1 - p) - ep \qquad (6.3)$$

where m defines the colonization rate of local populations and e the extinction rate of local populations. Therefore ep represents loss (extinction) of local populations from the whole metapopulation. The birth rate of local populations is represented by $mp(1 - p)$. The reason why p is multiplied by $1 - p$ can be conceptualized as a local neighbourhood problem. If there is one occupied patch surrounded by eight empty patches then the probability of colonization of any one empty patch is likely to be less than if there was one empty patch surrounded by eight occupied patches. Thus in determining the colonization probability the density of occupied patches and empty patches needs to be combined by multiplying them. In reality, the colonization (m) and extinction (e) parameters are likely to be complex functions of a series of variables. Thus m involves finding a new site, which depends on propagule dispersal (in turn dependent on the taxon and habitat under scrutiny and perhaps wind or water current speed or abundance of animal dispersers), the spatial distribution of occupied and unoccupied sites, initial establishment of propagules and subsequent population growth; whilst e is affected by the factors discussed in Chapter 2. (Indeed the methods of Foley (1994) described in Chapter 2 have been incorporated into metapopulations models, Hanski *et al.* 1996.)

Now let us consider the dynamics of the system described by Eqn 6.3. What are the conditions of metapopulation increase, no change or decline? No change in p is given by $dp/dt = 0$:

$$0 = mp(1 - p) - ep$$

$$p = 1 - e/m$$

Increase in p will occur if $dp/dt > 0$. Using the above rearrangement this leads to the inequality $p < 1 - e/m$. For decrease the required inequality is $p > 1 - e/m$.

In considering these results let us think about what is required of the meta-population model. If extinction and colonization rates (death and birth) are balanced, i.e. $e = m$, then there should be no change in metapopulation size. Similarly if the extinction rate (e) is greater than the colonization rate (m) then the metapopulation should decrease and increase when the colonization rate exceeds the extinction rate.

These conclusions only agree in part with the results of the manipulation of the equation. The problem is that when $dp/dt = 0$ we have the result $p = 1 - e/m$ (above). If $e = m$ then $p = 0$. Thus when extinction balances colonization we are left with the odd result that there are no local populations. If colonization is greater than extinction then e/m is less than 1 but greater than 0 and therefore $1 - e/m$ lies between 1 and 0 resulting in a positive value of p. Despite the drawbacks of Eqn 6.3 we will see in the next section how it can be used to explain limits to species range and how more complex models can give similar predictions. Hanski (1991) considers various refinements and developments of the basic Levins model.

6.2.3 Geographic range of plant species: the model of Carter and Prince

The plant metapopulation model of Carter and Prince (1981, 1988) linked the ideas of Levins with the one-equation models of infectious diseases described in Chapter 7 in an attempt to explain the geographical distribution limits of plant species. In particular they challenged the view that distribution was determined solely by correlation to climate variables, for example, that the northerly distribution limit of plant species in Britain was determined by physiological intolerance of cold winters (see examples in Carter & Prince 1988). Carter and Prince used a differential equation to describe a strategic model (see Chapter 1) of plant distribution:

$$\frac{dy}{dt} = bxy - cy \qquad (6.4)$$

where x = number of susceptible sites (sites available to be colonized), y = number of infective sites (occupied sites from which seed is produced and dispersed), b = infection rate and c = removal rate.

b and c are essentially local population birth (colonization) and death (extinction) rates and therefore equivalent to m and e in Levins equation (Eqn 6.3). Similarly x and y are related to $1-p$ and p in the same equation where p = proportion of local populations which are extant and potentially 'infective' and $1-p$ is the proportion of vacant (and therefore susceptible) sites.

Therefore any conclusions from the Carter and Prince model are relevant to metapopulations in general as defined by Levins. This illustrates how independent lines of enquiry about different ecological or biological systems may result in an appreciation of similar dynamic principles which can be described by the same equation or model. The important conclusion of Carter and Prince was that, along a climatic gradient, a very small change in, say, temperature, might tip the balance from metapopulation persistence to metapopulation extinction. In Carter and Prince's (1988) own words: 'a climatic factor might lead to distribution limits that are abrupt relative to the gradient in the factor, even though the physiological responses elicited might appear too small to explain such limits'. Thus climate (and physiological) factors are still important but their effects are amplified and made non-linear by the threshold properties of Eqn 6.3 or 6.4.

The results of these simple models are supported by the conclusions from more complex models. For example, the model of Herben *et al.* (1991) examined the dynamics of the moss *Orthodontium lineare* which occurs on tem-porary substrates such as rotting wood. This is a good candidate for metapopulation dynamics as persistence of the metapopulation requires dispersal from one local rotting wood population to another. The metapopulation is also known to be spreading in range. *Orthodontium lineare* is a native of the southern hemisphere but is now spreading rapidly through western and central Europe. Whilst the model of Herben *et al.* was intended to be more realistic than the strategic model of Carter and Prince, the same basic conclusions were reached. The model of Herben *et al.* included deterministic increase on occupied logs until carrying capacity was reached (recall that the Levins/Carter & Prince model does not have any local population dynamics other than zero to carrying capacity and vice versa), and the assumptions that dispersal by spores was in proportion to local population size and that spore dispersal distance declined exponentially from an occupied log. The results of this model supported the idea of Carter and Prince (and Levins) that there is a threshold for metapopulation persistence. In this case the percentage of logs oc-cupied was a non-linear function of probability of local population establishment (p_{est}, Fig. 6.4).

At a value of p_{est} of about 0.0002 the model predicted a sudden increase in the percentage of occupied sites. In other words there was a **threshold** value of p_{est} above which metapopulation persistence was likely to be high. If p_{est} is a function of climate then this would produce exactly the type of sharp break in species range predicted by Carter and Prince. It seems that such thresholds may be generated in a variety of ways. The next section (6.2.4) develops this theme with an alternative formulation for the metapopulation equation 6.3. Section 6.3 considers how linear changes in gap frequency in grassland can also generate a threshold for plant population abundance and Section 6.4 describes how diffusion processes can lead to thresholds.

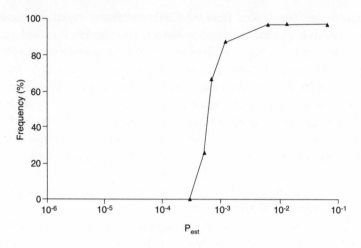

Fig. 6.4 Threshold in occupied sites with probability of establishment produced by model of Herben *et al.* (1991) after 100 simulation steps.

Question 6.2

How mathematically can the number of susceptible and infective sites, x and y (Eqn 6.4), be related to the proportion of empty and occupied sites $1-p$ and p (Eqn 6.3)?

6.2.4 An alternative metapopulation model using the logistic equation

In this section we use the logistic equation in place of Eqn 6.3 or 6.4 to model metapopulation dynamics and consider again the possibility of a threshold determining the edge of a species range. One key feature of the original model needs to be retained, i.e. that there is some interaction between the densities of infectives (occupied patches) and susceptibles (empty patches) in determining colonization rates. This interaction is represented by $p(1-p)$ in Eqn 6.3 and xy in Eqn 6.4. This can be taken further by considering the relationship between the relative or net colonization rate (m/e or $m-e$) and the density of susceptibles with respect to infectives (S/I). In the absence of any effect of S/I then the change in infectives (dI/dt) can be described as the net colonization rate multiplied by the number of susceptible patches:

$$\frac{dI}{dt} = (m-e)S \tag{6.5}$$

The relative colonization rate $m-e$ will be expected to vary with S and I. If S/I is high then $m - e$ should be low. If S/I is low then we expect $m - e$ to be close to its maximum value as most empty patches will be surrounded by occupied patches. There is a clear analogy with the logistic equation (Chapter 3). $m - e$ can be replaced by a value r (births minus deaths) and S/I by S'. The simplest reduction of r is linear with respect to S' which is described by $1-S'/K$. This is summarized by:

$$\frac{dI}{dt} = rS\left(1 - \frac{S'}{K}\right)$$

(6.6)

Note that, in contrast to the logistic equation, the rate of change variable (I) on the LHS is not the same as the variable on the RHS.

At equilibrium Eqn 6.6 is:

$$rS\left(1 - \frac{S'}{K}\right) = 0$$

So either $rS = 0$ which can be interpreted as $r = 0$ ($e = m$) and/or $S = 0$ (i.e. no remaining susceptible sites) or $1 - S'/K = 0$ and therefore $S' = K$ (as expected from the logistic equation). If $S' > K$ then $1 - S'/K$ is negative and so dI/dt is negative. Therefore I decreases and consequently S/I continues to increase. Thus K is a threshold condition for a ratio of S to I, i.e. S'. If S' is too high (above K) then the metapopulation cannot persist and the number of infectives declines (and so the proportion of susceptibles, S', increases).

In concluding the discussion of metapopulations we point the reader in the direction of model refinements: within-patch dynamics (mentioned for the models of Herben *et al.*) and consideration of patch connectivity. The former has been studied by Hastings and Wolin (1989) who assumed a general population growth function (e.g. the logistic) to describe population growth within a patch. Patch extinction and colonization rates were related to population size. Their model predicted a stable equilibrium distribution of patch sizes. They noted that this prediction is testable in the field when growth, colonization and extinction rates as functions of population size are known. Unfortunately there are few examples where such data are known. Fahrig and Merriam (1985) mixed theory and field data in an investigation of patch connectivity. They applied an interconnected habitat patch model to a population of white-footed mice inhabiting patches of forest in an agricultural landscape. Their prediction that mouse populations in isolated patches are more prone to extinction was supported by their field data.

6.3 Gap models and plant population thresholds

The work of Crawley and May (1987), Klinkhamer and De Jong (1989) and Silvertown and Smith (1989) has described an apparent threshold for plant population persistence determined by gap density in grassland (Fig. 6.5). They

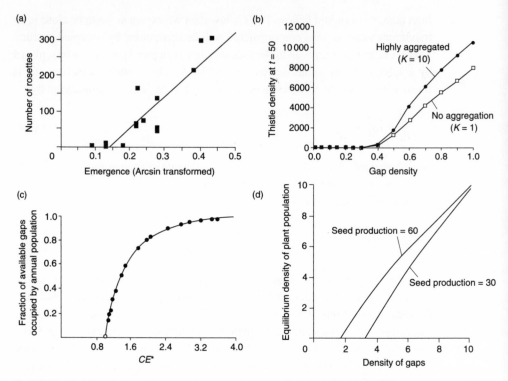

Fig. 6.5 Threshold for plant population persistence with a change in gap density. (a) Field data: the relationship between *Cirsium vulgare* rosette numbers and the probability of emergence (arcsin transformed) of seeds sown (from Silvertown & Smith, 1989); predictions of models: (b) Silvertown and Smith (1989); (c) Crawley and May (1987) CE^* is the product of annual fecundity (c) and proportion of gaps (E^*) at equilibrium; (d) Klinkhamer and De Jong (1989).

showed for short-lived herbaceous plants such as *Cirsium vulgare* that small changes in gap density due to, for example, disturbance by grazing animals or absence of perennials, resulted in large changes in the density of plants recruiting solely by seed. The recruitment of seeds into gaps in grassland was introduced in Chapter 4. In this section a simple model will be explored to explain this threshold and show how it can be related to the simulation models of Crawley and May and Silvertown and Smith and to the analytical model of Klinkhamer and De Jong. The threshold will be seen to arise directly from a spatially explicit model in which the seed are distributed in a particular way across a set of gaps.

Assume a field of area A is covered with n gaps of equal size (g); therefore the area of gaps = ng and the fraction of field covered by gaps (f) is:

$$f = \frac{ng}{A}$$

Now consider the proportion of gaps receiving one or more seeds, as only these seeds may be expected to germinate. Assume that a maximum of one seed can germinate per gap.

To estimate the proportion of gaps receiving one or more seeds assume a particular distribution of seeds amongst gaps, e.g. if seeds are distributed according to the Poisson (Box 6.1) distribution then the proportion of gaps containing no seeds is e^{-m} where m is the mean number of seeds per gap. Therefore the proportion of gaps containing one or more seeds is $1 - e^{-m}$. This was assumed by Crawley and May (1987) and Klinkhamer and De Jong (1989). Similarly one could assume a negative binomial distribution (Box 6.1). We use $k = 1$ which corresponds to a uniform distribution as used in Silvertown and Smith's simulation (Fig. 6.5b, the degree of aggregation did not affect the outcome). The effect of distributing seeds between gaps in this way, with a maximum of one survivor per gap, is to introduce density dependence into the model.

With the negative binomial and uniform distribution of seed the proportion of gaps containing one or more seeds is $m/(m + 1)$. m, the mean number of seeds per gap, is given by the total number of seeds (s) multiplied by the fraction falling into gaps divided by the number of gaps (n). We will assume that the fraction of seed falling into gaps is equivalent to the fraction of ground covered by gaps (f). Therefore:

$$m = \frac{sf}{n} \tag{6.7}$$

Now let us incorporate these details into a model of biennial population dynamics such as that described for *Cirsium vulgare* in Chapter 4 (a model of annual plant species could also be considered).

The number of first-year rosettes (R) in year $t + 1$ is given by the proportion of gaps which contain one or more seeds in year t ($m/m + 1$) multiplied by the number of gaps (n) and the probability of survival from seed to rosette (p_1):

$$R_{t+1} = \left(\frac{m}{m+1}\right)np_1 \tag{6.8}$$

Substitute the RHS of Eqn 6.7 for the numerator in Eqn 6.8 and cancel n:

$$R_{t+1} = \frac{s_t f p_1}{m+1} \tag{6.9}$$

Now assume that first-year rosettes survive with probability p_2 to become flowering plants (F) in the next year:

$$F_{t+2} = p_2 R_{t+1} \tag{6.10}$$

and that, in the same year, each flowering plant produces an average of q seeds.

Therefore, the total number of seeds (s) in year $t + 2$ is related to the number of rosettes in the previous year ($t + 1$):

$$S_{t+2} = qp_2 R_{t+1} \qquad (6.11)$$

Substitute the RHS of Eqn 6.9 into Eqn 6.11:

$$S_{t+2} = \left(\frac{s_t fp_1}{m+1} \right) qp_2$$

qp_1p_2 combines fecundity and survival and therefone can be represented by λ:

$$S_{t+2} = \frac{s_t f \lambda}{m+1} \qquad (6.12)$$

Equation 6.12 at equilibrium (s^*) is:

$$s^* = \frac{s^* f \lambda}{m+1}$$

Rearrange to make m the subject of the equation:

$$m = f \lambda - 1$$

Now substitute for m from Eqn 6.7:

$$\frac{s^* f}{n} = f\lambda - 1$$

$$s^* = n\lambda - \frac{n}{f} \qquad (6.13)$$

We know that $f = ng/A$ and therefore $n/f = A/g$.

A/g represents the ratio of the size of the field to the size of the gap and can be replaced by α giving a simpler version of Eqn 6.13:

$$s^* = n\lambda - \alpha \qquad (6.14)$$

This relationship between the equilibrium number of seeds (s^*) and the gap density (n) is shown in Fig. 6.6. From Eqn 6.14 and Fig. 6.6 the threshold effect of gap density can be seen. When $n < n_T$ then $s^* = 0$ (s^* can theoretically also be negative but this is not ecologically realistic). The analytical results of Eqn 6.14 therefore support the results of the simulation of Silvertown and Smith (1989). (Furthermore we could use field data to parameterize the model, e.g., similar to that in Fig. 6.5(a).

The model of Klinkhamer and De Jong (1989) reached similar conclusions showing that if $dsc < 1$ (where d is the density of gaps, s is the seed production and

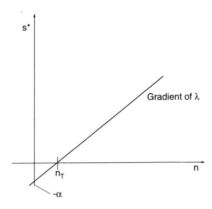

Fig. 6.6 Threshold predicted from λ and α using Eqn 6.14.

c is the gap area) the population would become extinct and if $dsc > 1$ then a population equilibrium would exist. This may begin to sound familiar. dsc is effectively a finite rate of increase for a gap-dependent grassland plant. The same is true of $n\lambda$ in Eqn 6.14. The threshold result is therefore as predicted by the discrete time density-independent and density-dependent models first introduced in Chapter 2, i.e. the requirement for a population to deterministically not go extinct and (with density dependence) to have a positive equilibrium population size is that $\lambda \geqslant 1$.

The results in this section are likely to be generally applicable where recruitment into gaps is important, e.g. seedlings of tree species into forest gaps or planktonic larvae of barnacles into rocky shore gaps.

6.4 Diffusion processes

6.4.1 Introduction
In the models of Sections 6.1–6.3 we have largely ignored the details of the dispersal phase, assuming it to be a simple function of the number of individuals in a local population. This is reasonable as a first approximation. Pragmatically, too little is usually known about dispersal to model it in anything other than the simplest way. One simple approximation which leads to interesting results is to assume that the individuals **diffuse** out from a source population. This may describe the expansion of an invading species across suitable habitat or the movement of individuals between local populations across uncolonizable habitat. Diffusion is a random and continuous process with each particle or individual going on a random walk from its source position. Whilst the concept is straightforward, diffusion models are complex because they require a method of summarizing all the random movements at each point in time.

6.4.2 *Applications*

Applications of diffusion models in ecology include the work of Morris (1993) on pollen dispersal and insect movement, Foley (1994) who used random walks to predict population extinction (Chapter 2) and marine ecologists studying the movement of algae in water bodies (see below). Segel (1984) provides an introduction to diffusion models of bacteria movement.

Diffusion can be described by the use of partial differential equations (PDEs, see Chapter 7). PDEs are needed because movement and/or abundance of individuals is dependent on two variables: spatial position and time. Maynard Smith (1968) provides an accessible introduction to partial differentiation applied to biological problems. He describes the diffusion of a substance along a tube. The change in concentration (x) with time (t) is related to the change in concentration with distance (s) according to the following PDE:

$$\frac{\partial x}{\partial t} = \mu \frac{\partial x^2}{\partial s^2}$$

where μ is a constant. This equation is well known in the mathematical literature as the one-dimensional heat equation. An ecological use of this equation is the dispersal of individuals along a linear route, e.g. plants dispersing along a roadside. In this case the constant might combine finite rate of increase and mean dispersal distance. A major problem is that only linear PDEs have analytical solutions. Despite this it is worth considering the results of PDE analyses of diffusion processes in ecology as they have produced results which support the results of boundaries/thresholds above. Also, as Maynard Smith observed (and this applies to many mathematical arguments), 'a familiarity with the notation enables one to follow other people's arguments, even if one could not have developed the argument oneself'.

A moderately simple model was used by Kierstead and Slobodkin (1953) who related the size of a plankton patch to the degree of turbulent diffusion (see Wyatt 1974 and Steele 1974). Kierstead and Slobodkin wanted to determine whether or not there is a minimum water mass size below which no increase in phytoplankton concentration is possible. They demonstrated the existence of a threshold condition for a patch of algae to cause a red tide. Steele (1974) also looked at diffusion in marine systems, again asking what causes 'patchiness' of algae in the sea. He considered that 'lateral turbulent diffusion of the water is a dominant physical process and can be expressed in a mathematical formulation'. Steele compared this physical process with the ecological process of herbivory using a pair of PDEs and showed that if the system without turbulence is unstable then diffusion under certain conditions could stabilize the system and, above a critical value, the system would destabilize again. Steele concluded that 'if an ecosystem is basically unstable when considered without diffusion processes then diffusion can remove the instability at smaller scales but not larger scales'.

6.5 Conclusion

Spatially explicit models of population and community dynamics are currently one of the most exciting growth areas in ecology. Their topicality aside, they contain several important messages for ecologists. First, they reveal how independent lines of enquiry can lead to similar ecological principles, e.g. the production of thresholds, and the importance of spatial scale. Second they have important implications for the relationship between theory and field ecology. For example, metapopulation models have demonstrated the clear requirement for detailed field study of colonization and extinction rates and associated parameters such as fragmentation of habitat. Finally, they caution us against simplistic interpretations of large spatial scale anthropogenic effects such as global climate change. The preponderance of threshold possibilities shows that we must expect a non-linear response to climate change. Species will not simply shift their range in linear procession; there will be extinctions and explosions, as predators and prey uncouple, metapopulation parameters are tweaked and finite rates of increase shift about their thresholds.

CHAPTER 7

Disease and biological control

In this final chapter we are going to return to models in continuous time, using disease in insect and human populations as the theme. Focusing on disease allows us to explore a progression of models; early models of host–pathogen interactions involved just one dynamic variable—describing the rate of spread of a disease through a constant population. These ideas were then expanded to include two, three or four dynamic variables, as the assumption of a constant population was relaxed, and various other components such as acquired immunity were included. We will cover how to analyse the stability of a model containing these dynamic variables, and discuss whether or not the behaviour of these models provides an explanation of the patterns we observe in the field.

In earlier chapters, field data have provided estimates of model parameters. In fact, the models themselves can aid in the design of field experiments. Insect–pathogen systems provide some examples of cases where models have been parameterized and then tested. One of these is discussed in Section 7.5.

7.1 A one-equation model of disease

7.1.1. The basic model
First, consider the spread of a flu virus through the human population. This is obviously suitable for a differential equation framework, as human generations overlap, and we can assume that the development of infection in individuals is not synchronized in any way. If we make the simplifying assumption that the population size is constant (for the time being), we need only concern ourselves with the change in the number of people carrying the virus. This can be expressed as follows:

$$\begin{matrix}\text{Change in} \\ \text{the number of} \\ \text{cases of flu}\end{matrix} = \begin{matrix}\text{The number of new} \\ \text{people infected}\end{matrix} - \begin{matrix}\text{The number of new deaths} \\ \text{and recoveries}\end{matrix} \qquad (7.1)$$

Note that the LHS of the equation concerns the change in the number of flu cases, not the total number of cases at any one time. If people become infected very easily, then the RHS, will be positive, and the number of new cases will increase rapidly. If individuals recovery very quickly before they have had a chance to pass it on to someone else, then the RHS will be negative, and the flu epidemic will be short-lived. Let us now consider each of the three terms in Eqn 7.1.

1   Change in the number of cases of flu. As in Chapter 3, we are talking about instantaneous rates of change, and therefore this is represented by dI/dt.

2   The number of new people infected. A new infection will occur each time a person carrying the virus comes into contact with a person who is susceptible to it. If we consider one infected individual in an otherwise healthy population, the number of contacts that individual will make per unit time is proportional to the population density (denoted by S, the density of susceptibles). As S increases, so will the contact rate between the one infected individual and the healthy individuals. Now if instead of one infected individual, there are I infected individuals, the contact rate becomes proportional to $S \times I$ per unit time (as in Eqns 6.3 and 6.4). However, not every contact will lead to a successful infection, so we need a constant of proportionality that reflects the ease with which the virus is transmitted from one person to another. If we call this **transmission parameter** β, then we have:

$$\beta SI \tag{7.2}$$

3   The number of new deaths and recoveries. Death may be due to infection or 'other reasons'. The number that die per unit time will be proportional to I. If we denote α and b as the instantaneous per capita death rates due to disease and other factors, respectively, then per unit time there will be $(\alpha + b)I$ new deaths. Similarly, if γ is the instantaneous recovery rate, then in the same small time interval γI individuals will have recovered. So the net losses from the infected portion of the population per unit time amount to:

$$(\alpha + b + \gamma)I \tag{7.3}$$

Combining equations 7.1–7.3, we can express the rate of change of flu infections in the population as:

$$\frac{dI}{dt} = \beta SI - (\alpha + b + \gamma)I \tag{7.4}$$

7.1.2 Persistence and spread of the disease

So what is the ultimate fate of the virus in the human population? Referring back to the simple one-population models of Chapter 3, it is necessary for Eqn 7.4 to be positive for the virus to spread and persist, i.e.

$$\beta SI > (\alpha + b + \gamma)I$$

$$\therefore \frac{\beta S}{(\alpha + b + \gamma)} > 1 \tag{7.5}$$

This provides us with a criterion for a disease to invade and establish in a population. However, is it possible to provide this algebraic expression with a biological interpretation? The subpopulation containing infected individuals is being depleted at a constant rate equal to $(\alpha + b + \gamma)$. Therefore the mean length of time an individual remains infected (before dying or recovering) is $1/(\alpha + b + \gamma)$. An infected individual causes βS new infections per unit time. Combining this information, one infection will cause, on average, $\beta S/(\alpha + b + \gamma)$ new infections before death or recovery. This quantity must be greater than 1 for the number of infections to increase (as stated in Eqn 7.5). Thus through biological reasoning, we have arrived at the same criterion. This expression (Eqn 7.5) is referred to as the **basic reproductive rate** of the disease, R_0, and provides the criterion for a disease to invade a population of susceptible individuals.

To derive an expression for the threshold density of susceptible people S_T required for the flu virus to invade an isolated population we need to rearrange Eqn 7.5 to make S the subject. Doing this we get:

$$S_T > \frac{(\alpha + b + \gamma)}{\beta} \tag{7.6}$$

This is the density of susceptible people required for the rate of acquisition of new infections to exceed the loss of infected individuals through death or recovery. If the number of susceptible humans falls below this threshold (e.g. an isolated island population), then the disease cannot persist within that population. Carter and Prince (1981) used similar arguments for thresholds of species range—see Chapter 6.

We are now reaching the limits of usefulness of this basic model. In summary, if $dI/dt > 0$ (or equivalently $R_0 > 1$) then the prevalence of the disease will increase. (Prevalence is defined as the proportion of the population carrying the disease.) Alternatively, if $dI/dt < 0$ (or $R_0 < 1$) then the disease will fade out. These conditions can also be expressed as a threshold population density that is required for the flu virus to persist in a population.

So far, it has been assumed that the host population size is constant: i.e. that births and deaths cancel each other out. However, if we consider a disease that is frequently fatal, then it may be very unrealistic to assume that the population is effectively constant. The disease itself may be a crucial factor causing population change. Now we extend the model by allowing the host population density to vary.

7.2 Biological control of insects using pathogens: a two-differential-equation model

7.2.1 The basic model

If we turn our attention from humans to insects, then violent population fluctuations are a more frequent observation in the literature! Many insect pests suffer from diseases caused by microparasites (e.g. viruses, bacteria, fungi, protozoa), and these may well be the cause of population change in some circumstances. These pathogens have been used successfully in a variety of pest control measures, as bioinsecticides, and also as longer-term biocontrol agents. If we extend our model to include change in the host population, we can then ask under what circumstances does the pathogen regulate the host? If we have a choice of pathogens, which would be most suitable as a biocontrol agent?

Anderson and May (1981) were the first to include the host population as a dynamic variable. This was a major step forward in the development of the theory underlying host-pathogen dynamics. To do this, we need to develop a second differential equation describing the change in the susceptible part of the host population.

The means by which individuals are gained and lost to the susceptible part of the host population are listed following the model developed by Anderson and May (1981: Model A).

Gains:

1   Reproduction by susceptible and infected individuals, producing new susceptible hosts,

2   Recovery of infected individuals.

Losses:

1   Death due to factors other than disease,

2   Infection of susceptible hosts, transferring them to the infected class.

Parameters for infection, death and recovery were defined in our previous model (Eqns 7.2 and 7.3). The only additional parameter we require is for reproduction. Let us assume that the *per capita* reproductive rate is the same for susceptible and infected individuals, and refer to this as *a*. We will also assume that the death rate due to causes other than disease (*b*) is the same for *S* and *I* individuals.

It is often useful to represent such a model as a flow diagram (Fig. 7.1).

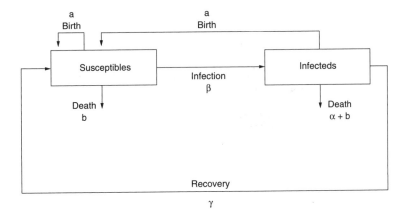

Fig. 7.1 A schematic representation of Model A (from Anderson & May 1981).

From Fig. 7.1, the second differential equation describing the instantaneous rate of change in susceptible individuals is as follows:

$$\frac{dS}{dt} = a(S+I) + \gamma I - bS - \beta SI \tag{7.7}$$

The total host population is $H = S + I$ (as opposed to $S + I$ = constant in our previous model).

If we use the same equation for dI/dt, we can write down the differential equation for the total host population H by summing the rates of change for the two components of H to obtain the overall rate of change for H itself. In doing this the expressions for infection and recovery disappear, as those terms just describe the transfer of individuals between categories within the host population. The end result is

$$\frac{dH}{dt} = rH - \alpha I \tag{7.8}$$

where $r = a - b$ = the intrinsic rate of increase for the host population (which we met in Chapter 3, Eqn 3.6). Whilst our model now consists of three equations, we only actually need two of them—the third can always be derived from the other two.

Question 7.1

What happens to the host population in the absence of disease?

7.2.2 Finding equilibrium

So in the absence of disease, the host population grows towards infinity. Under what conditions can the pathogen prevent such exponential growth? First, assume that the condition for the invasion of the pathogen into the host population ($R_0 > 1$) is upheld. Are there any circumstances under which the host and pathogen can coexist in a steady state? As before, we solve for equilibrium by setting dI/dt and dS/dt equal to zero. This provides us with three simultaneous equations, though we only need two of these (plus the identity $H = S + I$) to solve for S, I and H.

$$a(S+I) + \gamma I - bS - \beta SI = 0 \tag{7.9}$$

$$\beta SI - (\alpha + b + \gamma)I = 0 \tag{7.10}$$

$$rH - \alpha I = 0 \tag{7.11}$$

I and S can be determined at equilibrium (denoted by I^* and S^*) as follows. First, Eqn 7.10 is solved by cancelling I and rearranging to give $S^* = (\alpha + b + \gamma)/\beta$.

It is then simplest to substitute $S* + I* = H*$ into Eqn 7.11, and knowing $S*$ re-arrange to find $I* = r(\alpha + b + \gamma)/\beta(\alpha - r)$

So an equilibrium does exist at which both the host and pathogen are present, and the host population remains at

$$H* = \frac{\alpha(\alpha + b + \gamma)}{\beta(\alpha - r)} \tag{7.12}$$

For this to be a biologically meaningful solution, i.e. for the equilibrium to exist and be positive, it is necessary that $\alpha > r$.

If $R_0 > 1$ but $\alpha < r$ then the pathogen can still invade and persist within the host population. However, because $r > \alpha$ then $a > b + \alpha$, so the birth rate is greater than the death rate, and the population must continue to increase. The pathogen therefore does not have the ability to regulate the host population.

It is possible to be a little more precise about this. If we consider Eqn 7.8, as H becomes very large and the pathogen continues to increase in prevalence, then I will also become very large. At this limit, the difference between H and I becomes negligible and Eqn 7.8 can be rewritten approximately as:

$$\frac{dH}{dt} = (r - \alpha)H \tag{7.13}$$

In other words we have reduced the equation to the single-species exponential (Chapter 3, Eqn 3.6). H is therefore growing at a constant rate $r - \alpha$. This population is heading for infinity, albeit at a lesser rate than in the absence of disease.

7.2.3 Choosing the right control strategy

Pathogens which act as a source of mortality in an insect population may be manipulated as means of controlling that population. Pathogens such as viruses are particularly flexible in this respect, as they are suitable for use both as bioinsecticides and as classical biocontrol agents. Baculoviruses (a group of viruses that attack Lepidoptera) can either be formulated as sprays and so can be used in place of chemical insecticides, or introduced into a host population and allowed to spread naturally. These two different strategies may well call for different attributes on the part of the virus.

We have identified the conditions to be met for a pathogen to invade and persist in a host population. We have also shown that in this simplified system, there is a threshold virulence (α) required for the pathogen to be able to regulate the host. How does this help us with choosing a biocontrol agent? The attributes of the best control agent depend upon which control strategy we wish to adopt. If the aim is to maximize the speed with which the host is killed, then the biocontrol agent is being used as a biopesticide. Infected individuals have an expected lifespan of $1/(\alpha + b + \gamma)$. Therefore the fastest kill will be obtained with the pathogen of highest α, i.e. the most virulent pathogen. However, if the aim is to achieve a stable and low equilibrium density of the host population, the best strategy may require a different value of α. So our choice of pathogen for biological control will depend upon whether we aim to use it as a bioinsecticide when speed of kill will be of the essence, or establish it as a long-term biocontrol agent when host depression at equilibrium will be most important.

Question 7.2

Using Eqn 7.12, plot H^* as a function of α, fixing the other parameters as follows: $a = 2, b = 1, \gamma = 0.2, \beta = 0.02$. What sort of pathogen would you choose based on this information?

7.2.4 Stability analysis

We have informally looked at the behaviour of this model by inspecting the differential equations. In Chapter 3 we followed another informal method—phase plane analysis—when investigating the stability of the Lotka–Volterra model. Here we shall pursue a more analytical method of stability analysis using the Taylor series introduced in Chapter 5.

We have solved the two equations for equilibrium. The question remains as to how stable this equilibrium is. If there was a sudden influx of infected individuals into the population, would this cause the prevalence of the disease to increase on a permanent basis? Or would it decline back to equilibrium levels? In other words, how stable is this system of equations to small perturbations in the immediate vicinity of the equilibrium?

There are, in fact, two forms of stability. When disturbed, the population may decline smoothly to equilibrium, or overshoot and oscillate to either side of the equilibrium point with decreasing amplitude. Alternatively, it may oscillate to either side of the equilibrium indefinitely, with cycles of a particular amplitude, i.e. stable limit cycles (Chapter 2), and may be considered as a form of stability, because the population is neither becoming extinct, nor increasing without restraint.

Instability would involve the amplitude of any oscillations continuing to increase indefinitely (or else a smooth rise or decline in density). On the borderline between stability and instability is the special case of neutral stability (introduced in Chapter 3). This mathematical peculiarity is when a disturbance causes a population to remain fixed at its new position, or to oscillate around equilibrium with the amplitude of the oscillations being determined by the size of the original perturbation.

What we now require is a mathematical means of distinguishing between these different possibilities. The first step is to couch our disturbance in mathematical terms, i.e.

$$S_t = S^* + x_s \qquad (7.14)$$

where x_s is a small perturbation to the equilibrium population S^*, resulting in a new population density S_t.

We know that with the RHS of Eqn 7.7 rearranged:

$$\frac{dS}{dt} = (a-b)S + (a+\gamma)I - \beta SI$$

so

$$\frac{d(S^*+x_s)}{dt} = (a-b)S + (\alpha+\gamma)I - \beta SI \qquad (7.15)$$

Now the question becomes, does the magnitude of x_s increase or decrease with time? To simplify our reasoning, it is best to represent dS/dt as $f(S)$, i.e. a function of S. We can then make use of a Taylor approximation (Box 5.1) to find an expression for $f(S^* + x_s)$ which is easier to manipulate if we wish to evaluate a function at a particular value S which is known to be close to S^*, then

$$f(S) = f(S^*) + (S - S^*)f'(S^*) + \frac{1}{2!}(S - S^*)^2 f''(S^*) + \dots$$

where $f'(S)$ and $f''(S)$ are the first and second differentials of the function with respect to S, and $f(S^*)$ is the function evaluated when $S = S^*$. This makes is the evaluation of Eqn 7.15 much easier for three reasons:

1 $f(S^*)$ is equivalent to dS/dt at equilibrium, which is zero;
2 $S - S^* = x_s$, as defined above;
3 terms involving x_s^2 or higher can be ignored, as we have defined x_s as being small, so x_s raised to any power greater than 1 rapidly becomes too small to bother with.

Therefore, given these three results, Eqn 7.15 can be written as:

$$\frac{d(S^*+x_s)}{dt} \approx x_s f'(S^*) \text{ where } f'(S^*) = \frac{\partial\left[\frac{dS}{dt}\right]}{\partial S} \text{ evaluated at equilibrium.}$$

Whilst dS/dt = rate of change of S with time,

$$\frac{\partial\left[\frac{dS}{dt}\right]}{\partial S} = \text{rate of change of } \frac{dS}{dt} \text{ with } S.$$

This term is therefore a crucial factor which determines whether dS/dt will cause further increases or decreases in S as we perturb S itself.

Partial derivatives are covered in Box 5.1, and you should read this now if you have difficulty with the following.

Evaluate $\partial[\mathrm{d}S/\mathrm{d}t]/\partial S$ at equilibrium:

$$\frac{\partial\left[\dfrac{\mathrm{d}S}{\mathrm{d}t}\right]}{\partial S} = (a-b) - \beta I^* \quad \text{where} \quad I^* = \frac{r(\alpha+b+\gamma)}{\beta(\alpha-r)}$$

$$= \frac{r(a+\gamma)}{r-\alpha}$$

However, we should remember that changes in S do not occur in isolation. Such changes will also affect I, which in turn will instantly feedback to S (the feedback being instant due to the structure of the model). So we should also consider the rate of change of dS/dt with changes in I, i.e.

$$\frac{\partial\left[\dfrac{\mathrm{d}S}{\mathrm{d}t}\right]}{\partial S}$$

We should also repeat the whole process to consider how dI/dt is influenced by changes in both S and I. In total:

$$\frac{\mathrm{d}(S^* + x_s)}{\mathrm{d}t} \approx x_s \frac{\partial\left[\dfrac{\mathrm{d}S}{\mathrm{d}t}\right]}{\partial S} + x_I \frac{\partial\left[\dfrac{\mathrm{d}S}{\mathrm{d}t}\right]}{\partial I}$$

$$\frac{\mathrm{d}(I^* + x_I)}{\mathrm{d}t} \approx x_I \frac{\partial\left[\dfrac{\mathrm{d}I}{\mathrm{d}t}\right]}{\partial I} + x_s \frac{\partial\left[\dfrac{\mathrm{d}I}{\mathrm{d}t}\right]}{\partial S}$$

with the partial derivatives evaluated at equilibrium. This can also be expressed in matrix form:

$$\frac{\mathrm{d}X_t}{\mathrm{d}t} = \begin{bmatrix} \dfrac{\partial F_1}{\partial S} & \dfrac{\partial F_1}{\partial I} \\[2ex] \dfrac{\partial F_2}{\partial S} & \dfrac{\partial F_2}{\partial I} \end{bmatrix} \times \begin{bmatrix} x_S \\ x_I \end{bmatrix} \tag{7.16}$$

$$\begin{array}{l}\text{Rate of change of} \\ \text{perturbations}\end{array} = \text{Community matrix} \times \text{Initial perturbations}$$

where $F_1 = dS/dt$ and $F_2 = dI/dt$.

The square matrix containing the partial derivatives of our original model (evaluated at equilibrium) is often referred to as the community matrix (Section 5.2), because it encapsulates how each component of the model (in this case S and I—but these could equally well be two different species as in Section 5.2 and as we will see in Section 7.4) influences both itself and the other components. We know that the dominant eigenvalue of this matrix will summarize the behaviour of the perturbations over time—exactly the question we wish to answer.

The full community matrix is:

$$\begin{bmatrix} \dfrac{-r(a+\gamma)}{(\alpha-r)} & -(\alpha-r) \\ \dfrac{r(\alpha+b+\gamma)}{(\alpha-r)} & 0 \end{bmatrix}$$

The characteristic equation of this matrix (Chapter 4) is given by:

$$\sigma^2 + \frac{r(a+\gamma)}{\alpha-r}\sigma + r(\alpha+b+\gamma) = 0$$

The roots of this equation are the eigenvalues (denoted as σ in this example). We need not go as far as to solve the equation for σ to obtain the conditions for stability. As noted in Chapter 5, the perturbations must decrease with time if the equilibrium is to be stable. Therefore the dominant eigenvalue must be negative (and therefore both the eigenvalues must be negative). For this to be true, both the coefficients, $r(a+\gamma)/(\alpha-r)$ and $r(\alpha+b+\gamma)$, of the characteristic equation must be positive; which they will be if $\alpha > r$ and $r > 0$. Thus we have confirmed in mathematical terms that the equilibrium we found much earlier (Section 7.2.2) is a stable one.

Will the population exhibit a monotonic decline to equilibrium, or will it oscillate either side of the equilibrium with decreasing amplitude? Whilst the key to stability lies in the sign of the real part of the dominant eigenvalue, the answer to this question lies in the *imaginary* part of the eigenvalue. If there is no imaginary part to the eigenvalue, then the decline will be monotonic, otherwise the decline will involve oscillations of decreasing amplitude (see Fig. 7.2).

Even when it is known that $\alpha > r$, then the roots of the characteristic equation may or may not be complex, and this will depend upon the relative sizes of the model parameters (e.g. how great the difference is between α and r compared to the magnitude of γ). This distinction between monotonic or oscillating decline is of less dynamic consequence than that between stability and instability.

In this two-equation model, it was assumed that infection occurred through direct contact between infected and susceptible individuals. This is not the case

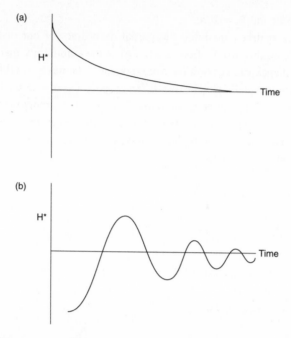

Fig. 7.2 Two populations disturbed from stable equilibria where the dominant eigenvalue has (a) no imaginary part; (b) an imaginary part.

with many diseases of insects, and now we shall move on to models that involve a different path of infection.

7.3 Modelling diseases with infective stages external to the host

Many pathogens that attack insects form persistent infective stages external to the host. The host then becomes infected upon consumption of these infective bodies. Baculoviruses, for example, form polyhedral inclusion bodies (PIBs) which can persist in the environment for many years. These consist of a protein coat surrounding a group of virions, the infectious units. When an insect consumes these PIBs, the protein coat breaks down, and the infection becomes established. When the insect dies, it releases many thousands of PIBs into the environment. Such a mode of infection requires us to include a third dynamic variable in our model—the density of PIBs in the environment. Such a model was first proposed by Anderson and May (1981), and is illustrated in Fig. 7.3.

The three differential equations describing the model illustrated in Fig. 7.3, using W to describe the density of PIBs in the environment are as follows:

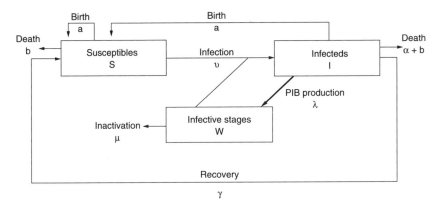

Fig. 7.3 A flow diagram of a three-variable model including external infective stages, from Anderson & May 1981; a = birth rate, α = death rate from disease, b = death rate from other causes, γ = recovery rate, υ = transmission rate, λ = rate of production of PIBs from an infected individual, μ = rate of inactivation of PIBs in the environment.

$$\frac{dS}{dt} = (a-b)S + (a+\gamma)I - \upsilon SW \qquad (7.17)$$

$$\frac{dI}{dt} = \upsilon SW - (\alpha + b + \gamma)I \qquad (7.18)$$

$$\frac{dW}{dt} = \lambda I - \left[\mu + \upsilon(S+I)\right]W \qquad (7.19)$$

The transmission parameter has been changed from β to v to emphasize that the infection process is very different from the 'direct contact' two-equation model. Once again v is a proportionality constant representing the transmission efficiency. However, the whole process of transmission has been broken into three components: λ, the rate of liberation of PIBs (or other infectious units) into the environment, μ, the rate at which these infectious units decay, and then v, the proportionality constant measuring the efficiency with which a new infection is established upon consumption of PIBs by the host. In the direct contact model, all three components were essentially embodied in the one parameter β.

As with the first model, R_0 needs to be greater than 1 for the infection to persist in the insect population; but how is R_0 defined for this model? This can be deduced from the model by considering the fate of a single PIB. The infection process occurs at rate υWH, i.e. per PIB it occurs at rate υH. Whilst in the external environment, PIBs are also decaying at rate μ. Each PIB will succumb to one of these two fates. Therefore the probability of any one PIB successfully causing a new infection rather than decaying is:

$$\frac{\upsilon H}{\mu + \upsilon H}$$

Once in a host, the infected host is expected to live for an average duration of

$$\frac{1}{(\alpha + b + \gamma)}$$

during which time it will be producing new PIBs at a rate λ. Consequently, in total, one PIB will be responsible for

$$\frac{\lambda}{(\alpha + b + \gamma)} \left(\frac{\upsilon H}{\mu + \upsilon H} \right) \tag{7.20}$$

new PIBs on average. This is R_0 for the three-equation model, and this must be greater than one for the infection to persist. We will gradually construct a graphical description of the dynamics of this model, allowing only two parameters to vary; namely α and λ. This will illustrate that this model can display some interesting behaviour not found in the two-equation model. First of all, from an inspection of R_0, it is possible to draw a line below which the pathogen cannot persist. Given that the second fraction in Eqn 7.20 will always be positive and less than or equal to one, then for the pathogen will definitely go extinct if:

$$\lambda < (\alpha + b + \gamma)$$

If the other parameters are fixed at $r = 1$, $b = 1$, $\gamma = 0$, $\mu = 0.02$, then a graph can be drawn with λ on the vertical axis, and α on the horizontal axis, showing in which region of parameter space the pathogen will become extinct (Fig. 7.4).

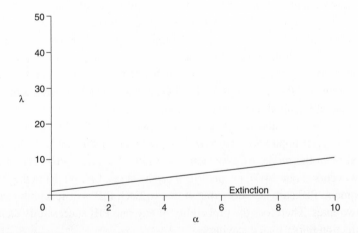

Fig. 7.4 The region of parameter space of the three-equation model in which the pathogen will become extinct given that $r = 1$, $b = 1$, $\gamma = 0$ and $\mu = 0.02$.

Further deductions can be made by inspecting the equilibrium values. The equilibrium expressions for S^*, I^* and W^* are:

$$I^* = \frac{\mu r(\alpha + b + \gamma)}{\upsilon[\lambda(\alpha - r) - \alpha(\alpha + b + \gamma)]}$$

$$S^* = \frac{\mu(\alpha - r)(\alpha + b + \gamma)}{\upsilon[\lambda(\alpha - r) - \alpha(\alpha + b + \gamma)]}$$

$$W^* = \frac{(\alpha + b + \gamma)r}{\upsilon(\alpha - r)}$$

It can now be seen that for there to be positive equilibria involving both I and S, the condition

$$\lambda > \frac{\alpha(\alpha + b + \gamma)}{(\alpha - r)} \tag{7.21}$$

must be fulfilled. So the rate at which PIBs are produced must exceed a certain level for a disease-regulated equilibrium to exist. This condition may then be added to the graph constructed earlier.

Given the parameter values above (Eqn 7.21), a line may be added to Fig. 7.4 to represent the division between parameter space in which pathogen–host equilibria may and may not exist. The line we wish to draw is expressed as:

$$\lambda = \frac{\alpha(\alpha + 1)}{\alpha - 1} \tag{7.22}$$

This is illustrated in Fig. 7.5.

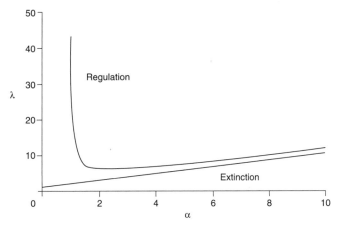

Fig. 7.5 Regions of parameter space for the three-equation model, where extinction of the pathogen will occur, and the boundary between regions where stable equilibria may and may not occur.

In fact it is possible to pinpoint the minimum on this curve by differentiation of Eqn 7.22, and setting the differential to zero. This gives a minimum at $\alpha = 2.41$, when $\lambda = 5.83$.

Below this boundary, but above the level at which the pathogen becomes extinct the pathogen will persist, but be unable to regulate the host population. As in the two-equation model, this will result in the host population increasing, but at a lower rate than in the complete absence of disease.

To complete the graph describing the dynamic behaviour of this model, we need to consider the nature of the regulated state. This can be done, as before, by producing the community matrix, and solving the characteristic equation. When the model involves three rather than two equations, this greatly increases the complexity of these operations—and it becomes rapidly necessary to use software to do these routine procedures for you. The characteristic equation for a three-equation model (Box 4.2) will take the general form

$$\sigma^3 + A\sigma^2 + B\sigma + C = 0$$

Rather than finding the roots of this equation (i.e. the eigenvalues), it is simply necessary to know that these roots have negative real parts (all must have negative real parts for the dominant eigenvalue to be negative). In fact, we can use a set of criteria (known as the Routh–Hurwitz stability criteria, Box 7.1) which state the conditions the coefficients must fulfil for this to be the case. For a cubic equation, these conditions are:

$$A > 0, \;\; B > 0, \;\; AB > C$$

When all of these conditions are fulfilled, the model is stable. Without going through the detailed algebra, in terms of λ and α we can now add a third line to the graph (defined by $AB = C$), enclosing the area in which the equilibria are stable (Fig. 7.6).

How does the model behave beyond this region of parameter space, i.e. at higher values of λ and α (the top right-hand corner of the graph)? This represents the region where one of the Routh–Hurwitz criteria is broken, namely $AB > C$. This results in the dominant eigenvalue having a positive real part, and therefore we do not get monotonic or oscillatory decline to equilibrium. Instead, initially at least, the population will rise or fall away from equilibrium. What happens ultimately cannot be deduced for certain from this local stability analysis. This is due to an assumption we made earlier in the stability analysis—when we used the Taylor expansion to find an approximate expression for $d(S^* + x_S)/dt$. As we were only considering the behaviour of the model close to S^*, we made the simplifying assumption that terms involving x_S^2 and higher order were not important in determining the direction of movement of the population when perturbed. As we move further away from S^*, these higher order terms will become increasingly important, and may (or may not) actually alter the direction of population change. So these non-linear terms

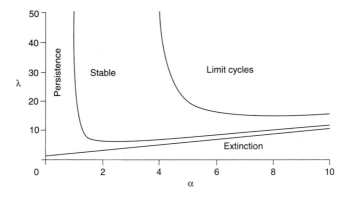

Fig. 7.6 Four regions of parameter space for the three-equation model.

will eventually decide whether the initial perturbations grow until the population reaches infinity (or extinction). In practice, numerical investigations are required to determine the effect of these non-linear terms; i.e. population trajectories should be simulated, after a small displacement from equilibrium, and followed for up to a few thousand generations (via computer simulations). If non-linear terms are crucial, then the system may settle down into stable limit cycles, i.e. populations of S, I and W cycling either side of their equilibria with an amplitude determined by parameter values. Indeed, this is what we find in this instance—the top right-hand area of the graph is a region of stable limit cycles. Given a particular set of parameter values, the amplitude of the stable limit cycles is fixed, regardless of the size of the initial perturbation. So even though one of the eigenvalues has a positive real part, the system is showing a form of stability due to the influence of non-linear terms.

When this model was first published (Anderson & May 1981), it caused much interest, as it illustrated that pathogens have the potential to cause cyclic dynamics in their host populations. This is particularly so for pathogens that have a high value of α (i.e. reasonably pathogenic) and low value of μ (i.e. persistent in the environment). This was coupled with the fact that many forest insects are observed to exhibit cyclic dynamics (with a period of several generations), and are known to support pathogens of moderate α and low μ. (Data on some forest insects and their pathogens are presented in Anderson & May 1981.)

Box 7.1 The Routh–Hurwitz criteria

In Chapter 4, the concept of eigenvalues was introduced. These are a set of numbers which can be used to represent a matrix. A 2 by 2 matrix will give rise to two eigenvalues, a 3 by 3 matrix three eigenvalues and so on. The dominant

Continued on page 176

Box 7.1 (*continued*)

eigenvalue is the largest of the set. In the process of deriving the eigenvalues from the matrix, the characteristic equation is produced. For a 2 by 2 matrix this will take the form

$$\sigma^2 + A\sigma + B = 0$$

This equation has two roots, and these are the two eigenvalues σ_1 and σ_2. In the stability analysis described in Section 7.3, it is not necessary to solve the equation to find the exact values of σ_1 and σ_2, it is simply sufficient to determine that these values are both less than zero. This means that the dominant eigenvalue will be less than zero, and any small perturbations will die out over time.

The above equation can be factored as follows:

$$(\sigma + C)(\sigma + D) = 0$$

where $C + D = A$ and $CD = B$. The eigenvalues are therefore $\sigma_1 = -C$ and $\sigma_2 = -D$. So both eigenvalues are negative if C and D are positive. Therefore we only need to know that the coefficients A and B are both positive, to deduce that the dominant eigenvalue for this matrix will be negative.

In summary, the Routh–Hurwitz criteria for a 2 by 2 matrix are that $A > 0$ and $B > 0$ where A and B are the coefficients of the equation $\sigma^2 + A\sigma + B = 0$.

This can be extended to state the stability criteria for larger matrices as summarized in the following table.

Table 1

Matrix size	Characteristic equation	R–H stability criteria
2×2	$\sigma^2 + A\sigma + B = 0$	$A > 0, B > 0$
3×3	$\sigma^3 + A\sigma^2 + B\sigma + C = 0$	$A > 0, C > 0$ $AB > C$
4×4	$\sigma^4 + A\sigma^3 + B\sigma^2 + C\sigma + D = 0$	$A > 0, C > 0, D \gg 0$ $ABC > C^2 + A^2D$
5×5	$\sigma^5 + A\sigma^4 + B\sigma^3 + C\sigma^2 + D\sigma + E = 0$	$A > 0, B > 0, C > 0, D > 0,$ $E > 0, ABC > C^2 + A^2D$ $(AD - E)(ABC - C^2 - A^2D) >$ $E(AB - C)^2 + AE^2$

It is not practicable to deal with larger matrices than these. For a fuller explanation see May (1974).

7.4 The special case of neutral stability

This method of stability analysis can yield one other class of result which will be well illustrated by referring back to a model already covered, namely the Lotka–Volterra model (Chapter 3). This is reproduced below for convenience.

$$\frac{dN}{dt} = r_1 N - \alpha PN$$

$$\frac{dP}{dt} = -r_2 P + \beta PN$$

Here we are using the version of the prey equation in which population growth is not constrained in the absence of predators, i.e. using the exponential rather than the logistic. It was this set of equations upon which Volterra based his original investigations.

Following the methods outlined in Section 7.2.4, we perform a stability analysis of this two-equation model. First, solving for equilibrium, we find:

$$N* = \frac{r_2}{\beta} \quad \text{and} \quad P* = \frac{r_1}{\alpha}$$

This provides the following community matrix:

$$\begin{bmatrix} 0 & \dfrac{-\alpha r_2}{\beta} \\ \dfrac{\beta r_1}{\alpha} & 0 \end{bmatrix}$$

which in turn gives the characteristic equation

$$\sigma^2 + r_1 r_2 = 0, \quad \text{and therefore} \quad \sigma = \pm \sqrt{-r_1 r_2} \ .$$

So the eigenvalues are pure imaginary numbers, the real part of the root being zero. What does this tell us about the behaviour of the model when disturbed from equilibrium? The imaginary part of the eigenvalues suggests that the behaviour will be oscillatory, whilst the real part suggests that the perturbations will neither increase nor decrease. In fact, the predator and prey populations will oscillate around their equilibria with constant amplitude, the size of which is determined by the size of the original perturbation.

Whilst oscillatory dynamics have been observed in the field, there are no examples of this specific behaviour—nor would we expect there to be, for the necessary conditions are so specific. First of all, the model would have to be an exact representation of what occurs in the field, and secondly, the parameters would have to return an eigenvalue the real part of which is exactly zero. The next section returns to the three-equation pathogen model, and explores how it can be fitted to field data.

7.5 Parameterizing and testing the insect–pathogen model

The model described by Eqns 7.17–7.19 is phrased in continuous time. This is obviously not appropriate for species which exist in temperate climates, for which events such as reproduction may be seasonal. However, such continuous models may be adapted to be appropriate within part of one season. If an insect species is univoltine, then within one season we do not have to consider reproduction, i.e. $a = 0$. Furthermore, in most insect–pathogen systems the recovery rate (γ) is zero. This immediately simplifies the model to:

$$\frac{dS}{dt} = -\upsilon SW - bS$$

$$\frac{dI}{dt} = -\upsilon SW - (\alpha + b)I$$

$$\frac{dW}{dt} = \lambda I - [\mu + \upsilon(S + I)]W$$

Dwyer and Elkinton (1993) further adapted this model to eliminate one of its more unrealistic features. When an insect becomes infected, there is a short period of time during which the virus is replicating within the insect's body cavity, and then the insect dies and liberates the PIBs into the environment. However, the death of a diseased insect is represented by the term $-\alpha I$, i.e. the rate at which infected hosts die is constant. If I_0 insects are infected at $t = 0$, we can plot a graph of I_t against time if (a) death occurs at a constant rate and (b) there is a fixed period of time before death (Fig. 7.7).

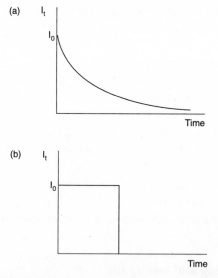

Fig. 7.7 (a) The decline in I_t with time if death occurs at a constant rate. (b) The decline in I_t with time if there is a fixed period of time before death.

Obviously the two graphs in Fig. 7.7 represent two extremes, but we may well expect the truth to be closer to (b) than (a). How may this be incorporated into the present model?

If we denote the length of time between infection and death as τ, then those infected insects that die at time t must have become infected at time $t - \tau$, when $vS_{(t-\tau)}W_{(t-\tau)}$ insects were infected. Therefore we can simply replace αI with $vS_{(t-\tau)}W_{(t-\tau)}$. These same insects are responsible for the production of PIBs at time t, so if we denote the number of PIBs produced when one caterpillar dies by Λ, then at time t, $\Lambda vS_{(t-\tau)}W_{(t-\tau)}$ PIBs are produced. This can therefore be substituted for the term λI. If we make two further simplifying assumptions, namely:

1 within the time-scale under consideration, b is negligible compared to α, and
2 the consumption of PIBs by the host insect does very little to actually deplete PIB density in the field,

we arrive at the fully reduced model used by Dwyer and Elkinton. This is

$$\frac{dS}{dt} = -vSW \tag{7.23}$$

$$\frac{dI}{dt} = vSW - vS_{(t-\tau)}W_{(t-\tau)} \tag{7.24}$$

$$\frac{dW}{dt} = \Lambda vW_{(t-\tau)}S_{(t-\tau)} - \mu W \tag{7.25}$$

Is this a reasonable representation of within season dynamics of the host and pathogen? This reduced model only requires four parameters, of which Λ and τ are easy to measure in the laboratory. μ and v are parameters that must be measured in the field.

In the absence of insect deaths, $dW/dt = -\mu W$. If the pathogen is introduced at density W_0 at time 0, and can be measured again as W_t at time t, it is possible to find an expression for μ.

$$\frac{dW}{dt} = -\mu W$$

To solve this equation we use integration which is effectively the reverse of differentiation:

$$\therefore \int_{W_0}^{W_t} \frac{1}{W} dW = \int_0^t -\mu dt$$

$$\therefore \mu = -\frac{1}{t} \ln\left[\frac{W_t}{W_0}\right]$$

In this way, we can calculate μ from field data, in the absence of any input into the PIB population. Similarly, if we consider the dynamics of susceptible and infected hosts over short periods of time (i.e. less than τ), we arrive at

$$\frac{dS}{dt} = -\upsilon SW \quad \text{and} \tag{7.26}$$

$$\frac{dI}{dt} = \upsilon SW \tag{7.27}$$

Furthermore, over such a short period, W_t may be taken as a constant (W_0)—though such assumptions should be verified during the course of an experiment.

Given the above assumptions, Eqn 7.27 can be solved to provide an expression for υ.

$$\frac{dS}{dt} = -\upsilon W_0 S_t$$

$$\therefore \int_{S_0}^{S_t} \frac{1}{S_t} dS = \int_0^t -v W_0 dt$$

$$\therefore \upsilon = -\frac{1}{tW_0} \ln\left[\frac{S_t}{S_0}\right] \quad \text{or equivalently}$$

$$\upsilon = -\frac{1}{tW_0} \ln\left[1 - \frac{I_t}{S_0}\right]$$

This provides an expression for υ involving terms that may be obtained from field sampling. A field trial to measure υ would involve releasing pathogen (W_0) and susceptible hosts (S_0) at time 0, and sampling at later timepoints to estimate the proportion infected (I_t/S_0) e.g., Goulson et al. (1995).

Once all model parameters have been estimated, it is possible to reproduce population trajectories for S and I within a season, and then compare these predictions with field data. Dwyer and Elkinton (1993) provide a rare and excellent example of this approach.

7.6 Conclusion

This chapter has followed the development of three basic models:
1   the first assuming a constant host population;
2   the second allowing the host population to vary;
3   the third including a dynamic variable which describes the density of infectious particles external to the host.

From these simple tactical models, we can derive conditions for the spread of a pathogen through a susceptible host population and predict when the pathogen could actually regulate that population. Since these models were first presented in the literature (Anderson & May 1981), they have been developed in a number of directions. For example, they have been adapted to include artificial manipulation of the pathogen by man, in an attempt to forecast the optimal control strategy (Anderson 1982). Certain simplifying assumptions, such as linear transmission, have been relaxed to investigate the impact this will have on the host and pathogen dynamics (e.g. Getz & Pickering 1983). Additional features have been added to the models, such as the existence of a reservoir into which the pathogen may enter (where it cannot infect hosts, but where mortality is low; Hochberg 1989). Indeed, this simple model framework has been the origin of a fertile area of theoretical and empirical research (for a review see Briggs *et al.* 1995). The challenge of the future is to strengthen the links between such theoretical studies and field data.

Answers to questions

Chapter 2

A2.1 $250/100 = 2.5$. By year 3 there will be $250 \times 2.5 = 625$ and by year 4 there will be $625 \times 2.5 = 1562.5$.

A2.2 (a) See Fig. A1 opposite this question. (b) The population with a λ of 2 increases in size geometrically, the population with a λ of 1 is constant in size over time whilst the population with a λ of 0.5 declines in size over time. (c) A population must have a value of λ equal to or greater than 1 to avoid extinction. (d) The λ of 2 is unrealistic over longer periods of time, because the population will reach enormous sizes and, theoretically, continue increasing *ad infinitum*. This finite rate of increase may be realistic in the short term, e.g. over a few years, when initial population densities are low. (e) The advantage of the logarithmic plot is that the change in population size over time is represented as a straight line, the rate of increase being indicated by the gradient of the line.

A2.3 There are several possibilities for improvement whilst retaining the basic design of the model. The number of trials for each initial population size (10) is low and could be increased. One could also try other initial population sizes, e.g. 100 and finally different pdfs could be used for λ.

A2.4 Not exactly because the results are dependent upon random events. However, if the number of trials for a given initial population size are sufficiently high, one would expect a similar answer for the probability of extinction. For example, if two model populations had both started with a population size of 5, one might have produced 65 out of 100 extinctions (within 20 years) and the other might have achieved 61 out of 100. In other words, if the exercise is highly replicated, then the answers from different sets of trials would converge.

A2.5 The randomly fluctuating time series (ai) corresponds to the Ricker–Moran plot (bi). This is because all combinations of N_{t+1} and N_t are possible in a randomly fluctuating population, represented by the scatter of dots in bi. In contrast, the chaotic dynamics in aii are produced by a deterministic model represented by a smooth curve through the points in bii. Thus each N_t has a unique corresponding value of N_{t+1}

(a)

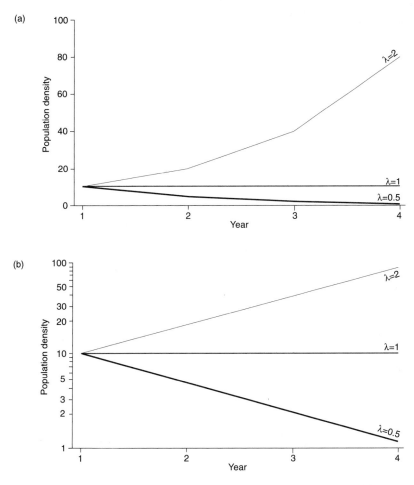

(b)

Fig. A1 Answers to Question 2.2. (a) Population density against time using Eqn 2.1 with finite rate of increase (λ) of 2, 1 and 0.5, respectively. (b) Previous graph replotted with logarithmic scale for density.

Chapter 3

A3.1 If the birth rate (b) exceeds the death rate (d) then $b > d$ (and $b - d$, or r, and hence dN/dt are positive) resulting in an increase in population size. Conversely, if $b < d$ then the population will decrease in size ($r < 0$, $dN/dt < 0$). As with the equivalent (density-independent) difference equation there is also an unstable equilibrium given by $b = d$ (so that $b - d = 0$ and there is no change in population size, i.e. $dN/dt = 0$. The unstable equilibrium is mathematically possible but ecologically extremely unlikely because, even if mean b did equal d, stochastic events would alter the values of b and d over time.

A3.2 Based on the rules for finding the derivative of a quotient (Box 3.1):

$$\frac{dN}{dt} = \frac{(1+e^{at})ab\,e^{at} - b\,e^{at}(a\,e^{at})}{(1+e^{at})^2}$$

$$\frac{dN}{dt} = \frac{ab\,e^{at}}{(1+e^{at})^2}$$

Substitute N for $b\,e^{at}/(1+e^{at})$:

$$\frac{dN}{dt} = aN\frac{1}{1+e^{at}} \tag{1}$$

Rearrange $N = b\,e^{at}/(1+e^{at})$ to give:

$$\frac{N}{b-N} = e^{at} \tag{2}$$

Substitute LHS of Eqn 2 for e^{at} in Eqn 1 and rearrange:

$$\frac{dN}{dt} = aN\left(\frac{b-N}{b}\right)$$

or

$$\frac{dN}{dt} = aN\left(1 - \frac{N}{b}\right)$$

In comparison to the logistic equation (Eqn 3.7) we see that $a = r$ and $b = K$. You could now find the derivative of Pearl and Reed's original three-parameter equation.

A3.3 Four equilibria (including $N = 0$) are produced by the sigmoidal change in harvesting. $N = 70$ is locally stable (as for $N = 80$ in Figs 3.7 and 3.8), $N = 30$ is locally unstable as perturbations either side of this population size will send the population towards $N = 10$ or 70. $N = 10$ is stable. Therefore this system has two locally stable equilibria. $N = 0$ is trivially stable in the sense that values of $N = 0$ will mean that $dN/dt = 0$ but any increase in N is predicted to result in increase towards $N_e = N_e\,e^{(0.34 - 0.0005N_e - 0.0018N_e)}$

A3.4 α indicates the efficiency with which predators remove prey, in other words, the loss of prey per unit density of predators. This will depend on the searching efficiency and attack rate of predators. Similarly β indicates the efficiency of conversion of prey into predators, i.e. the addition of new predators through birth per unit amount of prey. α is unlikely to equal β because the conversion of prey into predators is unlikely to be on a 1 : 1 basis. For example, it may take the consumption of many small mammals to produce a new predator such as a large cat or raptor. The exception is some solitary parasitoids where one parasitoid egg is laid on one host larva (see Chapter 6 for details of host–parasitoid interactions).

A3.5 Substitute values for r, a and b from Table 3.1 into Eqn 3.16 and set N_{t+1}, N_t and N_{t-1} equal to N_e:

$$N_e = N_e e^{(0.34 - 0.0005 N_e - 0.0018 N_e)}$$

$$1 = e^{(0.34 - 0.0023 N_e)}$$

Take natural logs:

$$0 = 0.34 - 0.0023 N_e$$

$$N_e = 147.8$$

Chapter 4

A4.1 The model population is expected to show either geometric increase, decrease or no change, dependent on the parameter values.

A4.2 We could examine data relevant to either the eigenvector or eigenvalue. If a sampling exercise was undertaken in the field in any one year and it was discovered that there were 4–5 times more rosettes than flowering plants this would support the eigenvalue result. Similarly if plant densities were followed over successive years, beginning at low population sizes (i.e. assuming density-independent conditions, see Chapter 2) the maximum rate of increase (dominant eigenvalue) could be checked. (We explore these tests of the model for a real set of data in Section 4.2.)

A4.3 q would be shifted from $s_{0,1}$ to the second equation:

$$R_{t+1} = f p_1 p_2 F_t$$

$$F_{t+1} = q s_{1,2} R_t$$

The calculation of the eigenvalue is unaffected, but the eigenvector changes:

$$\frac{R}{F} = \frac{\lambda}{q s_{1,2}}$$

i.e. divide all the previous values of R/F by q:

	Treatment							
	A	B	C	D	E	F	G	H
R/F model (original medium assumed)	2.80	2.98	2.72	5.40	0.93	1.57	2.48	5.13
q	0.084	0.084	0.246	0.246	0.084	0.084	0.246	0.246
R/F model (new, small assumed)	33.3	35.5	11.1	22.0	11.1	18.7	10.1	20.9

Chapter 5

A5.1 In the absence of competitors both Eqns 5.1 and 5.2 reduce to the logistic equation with equilibria of K_1 and K_2, respectively, i.e. both populations are assumed to be regulated by their own density which is also assumed for prey (Eqn 3.11). In contrast predators (Eqn 3.12) are assumed to exponentially decline in the absence of prey.

A5.2 The community matrix is based on three important assumptions:
1   that each population in the community has an equilibrium density,
2   that small perturbations of the populations occur near equilibrium;
3   that the behaviour of the populations near equilibrium can be described by linearized versions of Lotka–Volterra type equations.

A5.3 Begin by converting the percentages into probabilities (fractions):

$$\begin{pmatrix} 0.05 & 0.36 & 0.5 & 0.09 \\ 0.01 & 0.57 & 0.25 & 0.17 \\ 0 & 0.14 & 0.55 & 0.31 \\ 0 & 0.01 & 0.03 & 0.96 \end{pmatrix}$$

Then transpose the matrix (put row 1 in column 1, row 2 in column 2 and so on) and multiply by the column vector:

$$\begin{pmatrix} 10 \\ 10 \\ 10 \\ 10 \end{pmatrix}$$

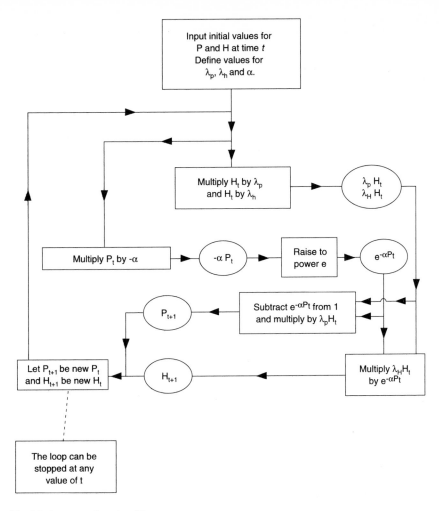

Fig. A2 Answer to Question 6.1.

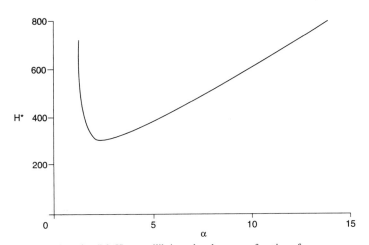

Fig. A3 Answer to Question 7.2. Host equilibrium abundance as a function of α.

This gives:

$$\begin{pmatrix} 0.5+0.1 \\ 3.6+5.7+1.4+0.1 \\ 5+2.5+5.5+0.3 \\ 0.9+1.7+3.1+9.6 \end{pmatrix} = \begin{pmatrix} 0.6 \\ 10.8 \\ 13.3 \\ 15.3 \end{pmatrix}$$

Note that the results in Table 5.4(b) assume a succession with different initial conditions.

Chapter 6
A6.1 (see Fig. A2, page 187)

A6.2 If $x + y$ is assumed to be the total number of sites then $y/(x + y)$ is p and $((x + y)/(x + y)) - y/(x + y)$, i.e. $1 - p$, equals $x/(x + y)$, which is the proportion of susceptible sites.

Chapter 7
A7.1 In the absence of disease, $I = 0$ and therefore the host population increases exponentially (if $a > b$; see Question 3.1).

A7.2 The graph on page 187 (Fig. A3) illustrates that the greatest host depression at equilibrium is achieved with an intermediate value of α.

Glossary of mathematical symbols and terms

Only those symbols which are used widely in this text and elsewhere are given here. Only terms which are not defined in the text are given here.

asymptote   A function value which is approached increasingly slowly with increasing or decreasing values of variables in the function. Thus $y = 1/x$ approaches the value 0 as x increases (but y never reaches 0).

coefficient   A value associated with a variable in a *function*, e.g. 2 and 3 are the coefficients in $y = 2x + 3x^2$.

complex number   A number involving the square root of a negative number. Based on the square root of -1, written as i (imaginary) or j. A complex number has a real part and an imaginary part, e.g. $3 + 2i$.

dN/dt   The rate of change of a variable (e.g. N) with change in a second variable (e.g. t). Also known as the derivative of N with respect to t.

$f(x)$ See *function*.

function y is a function of x if the value(s) of y is/are given when the value of x is given, denoted by $y = f(x)$.

integer   A whole number.

imaginary number   See *complex number*.

irrational numbers   Infinite non-recurring decimals (e.g. pi, $\sqrt{2}$).

K   Carrying capacity.

λ Finite rate of population change, also used to denote eigenvalue(s).

LHS   Left-hand side of equation.

log   Logarithm, assumed to be to base 10 unless otherwise stated.

ln   Natural logarithm, logarithm to the base e (see Box 2.2).

magnitude   Value of a real number regardless of sign (the magnitude of -2 equals the magnitude of 2).

N   Number or density of individuals. Subscript can give the time period $(t, t + 1)$, spatial location $(s, s + 1)$, species (i, j) or maximum value (max). Superscript * indicates equilibrium. Without subscripts N is still assumed to be at a particular location at a particular time.

parameter   Variable held constant for a particular example.

r   Intrinsic rate of population change $(r = \ln(\lambda))$.

rational number   Positive and negative integers, zero, fractions.

real number   Rational and irrational numbers.

reciprocal   The reciprocal of a number is 1 divided by that number.

RHS   right-hand side of an equation.

Σ   Summation sign, e.g.

$$\sum\nolimits_{i=1}^{s} N_i$$

means 'add up the values of N from N_1 to N_s'.

t   Time.

v   Population structure or age/size/stage-distribution vector (column matrix) giving number of individuals at different ages, sizes or stages.

References

Allee, W.C., Emerson, A.E., Park, O., Park, T. & Schmidt, K.P. (1949) *Principles of Animal Ecology.* W.B. Saunders, Philadelphia.

Anderson, R.M. (1982) Theoretical basis for the use of pathogens as biological control agents of pest species. *Parasitology*, **84**, 3–33.

Anderson, R.M. & May, R.M. (1981) The population dynamics of microparasites and their invertebrate hosts. *Philosophical Transactions of the Royal Society, B*, **291**, 451–524.

Atkinson, W.D. & Shorrocks, B. (1981) Competition on a divided and ephemeral resource: a simulation model. *Journal of Animal Ecology*, **50**, 461–71.

Baltensweiler, W. (1993) Why the larch bud-moth cycle collapsed in the subalpine larch–cembran pine forests in the year 1990 for the first time since 1850. *Oecologia*, **94**, 62–6.

Beddington, J.R. (1979) Harvesting and population dynamics. In: *Population Dynamics* (ed. R.M. Anderson, B.D. Turner & L.R. Taylor), pp. 307–20. Blackwell Scientific Publications, Oxford.

Beddington, J.R., Free, C.A. & Lawton, J.H. (1975) Dynamic complexity in predator–prey models framed in difference equations. *Nature*, **255**, 58–60.

Bernardelli, H. (1941) Population waves. *Journal of the Burma Research Society*, **31**, 1–18.

Berryman, A.A. (1992) On choosing models for describing and analyzing ecological time series. *Ecology*, **73**, 694–8.

Beverton, R.J.H. & Holt, S.J. (1957) On the dynamics of exploited fish populations. *Fishery Investigations. London Series*, **2(19)**, 1–533.

Birch, L.C., Park, T. & Frank, M.B. (1951). The effect of intraspecies and interspecies competition on the fecundity of two species of flour beetles. *Evolution*, **5**, 116–132.

Boorman, S.A. & Levitt, P.R. (1973) Group selection at the boundary of a stable population. *Theoretical Population Biology*, **4**, 85–128.

Briggs, C.J., Hails, R.S., Barlow, N.D. & Godfray, H.C.J. (1995) The dynamics of insect–pathogen interactions. In: *Ecology of Infectious Diseases in Natural Populations* (eds B.T. Grenfell & A.P. Dobson), 478–509 Cambridge University Press, Cambridge.

Broekhuizen, N., Evans, H.F. & Hassell, M.P. (1993) Site characteristics and the population dynamics of the pine looper moth. *Journal of Animal Ecology*, **62**, 511–18.

Bullock, J.M., Clear Hill, B. & Silvertown, J. (1994) Demography of *Cirsium vulgare* in a grazing experiment. *Journal of Ecology*, **82**, 101–111.

Burnett, T. (1958) A model of host-parasite interaction. *Proceedings of the 10th Internatinal Congress of Entomology*, **2**, 679–86.

Callaway, R.M. & Davis, F.W. (1993) Vegetation dynamics, fire and the physical environment in coastal central California. *Ecology*, **74**, 1567–78.

Carter, R.N. & Prince, S.D. (1981) Epidemic models used to explain biogeographical distribution limits. *Nature*, **293**, 644–5.

Carter, R.N. & Prince, S.D. (1988) Distribution limits from a demographic viewpoint. In: *Plant Population Ecology* (eds A.J. Davy, M.J. Hutchings & A.R. Watkinson), pp. 165–84. Blackwell Scientific Publications, Oxford.

Caswell, H. (1989) *Matrix Population Models.* Sinauer Associates, Sunderland, Mass.

Caughley, G. (1977) *Analysis of Vertebrate Populations.* Wiley, New York.

Colasanti, R.L. & Grime, J.P. (1993) Resource dynamics and vegetation processes: a deterministic model using two dimensional cellular automata. *Functional Ecology*, **7**, 169–76.

Collar, N.J. & Andrew, P. (1988). Birds to watch. The *ICBP World check–list of threatened birds*. ICBP, Cambridge.

Comins, H.N., Hassell, M.P. & May, R.M. (1992) The spatial dynamics of host–parasitoid systems. *Journal of Animal Ecology*, **61**, 735–48.

Crawley, M.J. (1992) Population dynamics of natural enemies and their prey. In: *Natural Enemies* (ed. M.J. Crawley), pp. 40–89. Blackwell Scientific Publications, Oxford.

Crawley, M.J. & Gillman, M.P. (1989) Population dynamics of cinnabar moth and ragwort in grassland. *Journal of Animal Ecology*, **58**, 1035–1050.

Crawley, M.J. & May, R.M. (1987) Population dynamics and plant community structure: competition between annuals and perennials. *Journal of Theoretical Biology*, **125**, 475–89.

Crawley, M.J., Hails, R.S., Rees, M., Kohn, D. & Buxton, J. (1993) Ecology of transgenic oilseed rape in natural habitats. *Nature*, **363**, 620–3.

Crombie, A.C. (1945) On competition between different species of granivorous insects. *Proceedings of the Royal Society of London, B*, **132**, 362–95.

Crombie, A.C. (1946) Further experiments on insect competition. *Proceedings of the Royal Society of London, B*, **133**, 76–109.

Crombie, A.C. (1947) Interspecific competition. *Journal of Animal Ecology*, **16**, 44–73.

Dempster, J.P. & Lakhani, K.H. (1979) A population model for cinnabar moth and its food plant, ragwort. *Journal of Animal Ecology*, **48**, 143–63.

Dempster, J.P., Atkinson, D.A. & French, M.C. (1995) The spatial dynamics of insects exploiting a patchy food resource. II. Movements between patches. *Oecologia*, **104**, 354–62.

Dwyer, G. & Elkinton, J.S. (1993) Using simple models to predict virus epizootics in gypsy moth populations. *Journal of Animal Ecology*, **62**, 1–11.

Elliott, J.M. (1977) Some methods for the statistical analysis of benthic invertebrates. 2nd edition. *Freshwater Biological Association Scientific Publication*, **25**. Titus Wilson and Son, Kendal.

Elliott, J.M. (1994) *Quantitative Ecology and the Brown Trout*. Oxford Series in Ecology and Evolution. Oxford University Press, Oxford.

Ellner, S. & Turchin, P. (1995) Chaos in a noisy world: new methods and evidence from time–series analysis. *American Naturalist*, **245**, 343–75.

Elton, C.S. (1958) *The Ecology of Invasions by Animals and Plants*. Methuen, London.

Elton, C. & Nicholson, M. (1942) The ten year cycle in numbers of the Lynx in Canada. *Journal of Animal Ecology*, **11**, 215–44.

Fahrig, L. & Merriam, G. (1985) Habitat patch connectivity and population survival. *Ecology*, **66**, 1762–8.

Foley, P. (1994) Predicting extinction times from environmental stochasticity and carrying capacity. *Conservation Biology*, **8**, 124–37.

Frelich, L.E., Calcote, R.R., Davis, M.B. & Pastor, J. (1993) Patch formation and maintenance in old growth hemlock–hardwood forest. *Ecology*, **74**, 513–27.

Gardner, M.R. & Ashby, W.R. (1970) Connectance of large dynamic (cybernetic) systems: critical values for stability. *Nature*, **228**, 784.

Gause, G.F. (1932) Experimental studies on the struggle for existence. I. Mixed populations of two species of yeast. *Journal of Experimental Biology*, **9**, 389–402.

Gause, G.F. (1934) *The Struggle for Existence*. Williams & Wilkins, Baltimore. Reprinted (1964) Hafner, New York.

Gause, G.F. (1935) Experimental demonstration of Volterra's periodic oscillation in the number of animals. *Journal of Experimental Biology*, **12**, 44–8.

Getz, W.M. & Pickering, J. (1983) Epidemic models: thresholds and population regulation. *American Naturalist*, **121**, 892–8.

Gillman, M.P. & Crawley, M.J. (1990) A comparative evaluation of models of cinnabar moth dynamics. *Oecologia*, **82**, 437–45.

Gillman, M.P. & Dodd, M. (in press) The variability of orchid populations. *Botanical Journal of the Linnean Society*.

Gillman, M.P. & Silvertown, J. (1997) Population extinction and the uncertainty of measurement. *Proceedings of the JNCC/BES Symposium on the role of genetics in conserving small populations*.

Gillman, M.P., Bullock, J.M., Silvertown, J. & Clear Hill, B. (1993) A density dependent model of

Cirsium vulgare population dynamics using field estimated parameter values. *Oecologia*, **96**, 282–9.

Goodman, D. (1987) the demography of chance extinction. In: *Viable Populations for Conservation* (ed. M.E. Soule), pp. 11–34. Cambridge University Press, Cambridge.

Hallett, J.G. (1991) The structure and stability of small mammal faunas. *Oecologia*, **88**, 383–93.

Hanski, I. (1991) Single-species metapopulation dynamics: concepts, models and observations. *Biological Journal of the Linnean Society*, **42**, 17–38.

Hanski, I. & Gilpin, M. (1991) Metapopulation dynamics: brief history and conceptual domain. *Biological Journal of the Linnean Society*, **42**, 3–16.

Hanski, I., Foley, P. & Hassell, M. (1996) Random walks in a metapopulation: how much density dependence is neccessary for long–term persistence? *Journal of Animal Ecology*, **65**, 274–82.

Hanski, I. & Gyllenberg, M. (1993) Two general metapopulation models and the core–satellite species hypothesis. *American Naturalist*, **142**, 17–41.

Harrison, S. (1991) Local extinction in a metapopulation context: an empirical evaluation. *Biological Journal of the Linnean Society*, **42**, 73–88.

Harrison, S. (1994) *Metapopulations and conservation in large-scale Ecology and Conservation Biology* (eds P.J. Edwards, R.M. May & N.R. Webb), pp. 197–237, Blackwell Science, Oxford.

Harrison, S., Quinn, J.F., Baughman, J.F., Murphy, D.D. & Ehrlich, P.R. (1991) Estimating the effects of scientific study on two butterfly populations. *American Naturalist*, **137**, 227–43.

Hassell, M.P. (1975) Density dependence in single species populations. *Journal of Animal Ecology*, **42**, 693–726.

Hassell, M.P. (1976) *The Dynamics of Competition and Predation*. Arnold, London.

Hassell, M.P. & Comins, H.N. (1976) Discrete time models for two-species competition. *Theoretical Population Biology*, **9**, 202–21.

Hassell, M.P. & Godfray, H.C.J. (1992) The population biology of insect parasitoids. In: *Natural Enemies* (ed. M.J. Crawley), pp. 265–92. Blackwell Scientific Publications, Oxford.

Hassell, M.P. & May, R.M. (1973) Stability in insect host–parasite models. *Journal of Animal Ecology*, **42**, 693–726.

Hassell, M.P. & Varley, G.C. (1969) New inductive population model for insect parasites and its bearing on biological control. *Nature*, **223**, 1133–7.

Hassell, M.P., Lawton, J.H. & May, R.M. (1976) Patterns of dynamical behaviour in single species populations. *Journal of Animal Ecology*, **45**, 471–86.

Hastings, A. & Wolin, C.L. (1989) Within-patch dynamics in a metapopulation. *Ecology*, **70**, 1261–6.

Herben, T., Rydin, H. & Soderstrom, L. (1991) Spore establishment probability and the persistence of the fugitive invading moss, *Orthodontium lineare*: a spatial simulation model. *Oikos*, **60**, 215–21.

Hochberg, M.E. (1989) The potential role of pathogens in biological control. *Nature*, **337**, 262–65.

Holling, C.S. (1966) the strategy of building models in complex ecological systems. In: *Systems Analysis in Ecology* (ed. K.E.F. Watt). 195–214, Academic Press, New York.

Hoppensteadt, C. (1982) *Mathematical Models of Population Biology*. Cambridge University Press, Cambridge.

Horn, H.S. (1975) Markovian properties of forest succession. In: *Ecology and Evolution of Communities* (eds M.L. Cody & J.M. Diamond), pp. 196–211. Harvard University Press, Cambridge, Mass.

Horn, H.S. (1981) Succession. In: *Theoretical Ecology* (ed. R.M. May), pp. 253–71. Blackwell Scientific Publications, Oxford.

Huffaker, C.B. (1958) Experimental studies on predation. Dispersion factors and predator–prey oscillations. *Hilgardia*, **27**, 343–83.

Hutchinson, G.E. (1948) Circular causal systems in ecology. *Annals of New York Academy of Science*, **50**, 221–46.

Hutchinson, G.E. (1978) *An Introduction to Population Ecology*. Yale University Press, New Haven and London.

Jones, T.H., Godfray, H.C.J. & Hassell, M.P. (1996) Relative movement patterns of a tephritid fly and its parasitoid wasps. *Oecologia*, **106**, 317–24.

Judson, O.P. (1994) The rise of the individual-based model in ecology. *Trends in Ecology and Evolution*, **9**, 9–14.

Kierstead, H. & Slobodkin, L.B. (1953) The size of water masses containing algal blooms. *Journal of Marine Research*, **12**, 141–7.

Kingsland, S.E. (1985) *Modeling Nature. Episodes in the History of Population Ecology*. University of Chicago Press, Chicago.

Klinkhamer, P.G.L. & De Jong T.J. (1989) A deterministic model to study the importance of density dependence for regulation and the outcome of intra-specific competition in populations of sparse plants. *Acta Botanica Neerlandica*, **38**, 57–65.

Krebs, C.J. (1994) *Ecology*. Harper Collins, New York.

de Kroon, H., Plaisier, A., van Groenendael, J. & Caswell, H. (1986) Elasticity: the relative contributions of demographic parameters to population growth rate. *Ecology*, **67**, 1427–31.

de Kroon, H., Plaisier, A. & van Groenendael, J. (1987) Density-dependent simulation of the population dynamics of a perennial grassland species, *Hypochaeris radicata*. *Oikos*, **50**, 3–12.

Lefkovitch, L.P. (1965) The study of population growth in organisms grouped by stages. *Biometrics*, **21**, 1–18.

Lefkovitch, L.P. (1967) A theoretical evaluation of population growth after removing individuals from some age groups. *Bulletin of Entomological Research*, **57**, 437–45.

Leslie, P.H. (1945) On the uses of matrices in certain population mathematics. *Biometrika*, **33**, 182–212.

Leslie, P.H. (1948) Some further notes on the use of matrices in population mathematics. *Biometrika*, **35**, 213–45.

Levins, R. (1966) The strategy of model building in population biology. *American Scientist*, **54**, 421–31.

Levins, R. (1968) *Evolution in Changing Environments*. Princeton University Press, Princeton, New Jersey.

Levins, R. (1969) Some demographic and genetic consequences of environmental heterogeneity for biological control. *Bulletin of the Entomological Society of America*, **15**, 237–40.

Levins, R. (1970) Extinction. In: *Some Mathematical Problems in Biology* (ed. M. Gerstenhaber), pp. 77–107. Mathematical Society, Providence, Rhode Island.

Levins, R. & Culver, D. (1971) Regional coexistence of species and competition between rare species. *Proceedings of the National Academy of Science*, **68**, 1246–8.

Lewis, E.G. (1942) On the generation and growth of a population. *Sankhya*, **6**, 93–6.

Lotka, A.J. (1925) *Elements of Physical Biology*. Williams & Wilkins, Baltimore. Reprinted 1956 by Dover Publications, New York.

Lotka, A.J. (1927) Fluctuations in the abundance of species considered mathematically (with comment by V. Volterra). *Nature*, **119**, 12–13.

McCarthy, M.A. (1996) Red kangaroo (*Macropus rufus*) dynamics: effects of rainfall, density dependence, harvesting and environmental stochasticity. *Journal of Applied Ecology*, **33**, 45–53.

McCauley, E., Wilson, W.G. & de Roos, A.M. (1993) Dynamics of age-structured and spatially structured predator–prey interactions: individual-based models and population-level formulations. *American Naturalist*, **142**, 412–42.

Mace, G.M. & Lande, R. (1991) Assessing extinction threats: towards a reevaluation of IUCN threatened species categories. *Conservation Biology*, **5**, 148–57.

Malthus, T.R. (1798) *An Essay on the Principle of Population*. J. Johnson London. Reprinted by Macmillan, New York.

Manly, B.J. (1990) *Stage-structured Populations. Sampling, Analysis and Simulation*. Chapman & Hall, London.

May, R.M. (1972) Will a large complex system be stable? *Nature*, **238**, 413–14.

May, R.M. (1973a) *Stability and Complexity in Model Ecosystems*. Princeton University Press, Princeton, New Jersey.

May, R.M. (1973b). On relationships among various types of population model. *American Naturalist*, **107**, 46–57.

May, R.M. (1974) Biological populations with nonoverlapping generations: stable points, stable cycles and chaos. *Science*, **186**, 645–7.

May, R.M. (1976) Simple mathematical models with very complicated dynamics. *Nature*, **261**, 459–67.

May, R.M. (1978) Host–parasitoid systems in patchy environments, a phenomenological model. *Journal of Animal Ecology*, **47**, 833–43.

May, R.M. (1981) *Theoretical Ecology. Principles and Applications*. Blackwell Scientific Publications, Oxford.

May, R.M. (1984) An overview: real and apparent patterns in community structure. In: *Ecological Communities: Conceptual Issues and the Evidence* (eds D.R. Strong, D. Simberloff, C.G. Abele & A.B. Thistle), pp. 3–18. Princeton University Press, Princeton, New Jersey.

May, R.M. & Oster, G.F. (1976) Bifurcations and dynamic complexity in simple ecological models. *American Naturalist*, **110**, 573–99.

May, R.M. & Watts, C.H. (1992) The dynamics of predator–prey and resource–harvester systems. In: *Natural Enemies* (ed. M.J. Crawley), pp. 431–57. Blackwell Scientific Publications, Oxford.

May, R.M., Conway, G.R., Hassell, M.P. & Southwood, T.R.E. (1974) Time delays, density dependence and single species oscillations. *Journal of Animal Ecology*, **43**, 747–70.

Maynard Smith, J. (1968) *Mathematical Ideas in Biology*. Cambridge University Press, Cambridge.

Maynard Smith, J. (1974) *Models in Ecology*. Cambridge University Press, Cambridge.

van der Meijden, E., van Wijk, C.A.M. & Kooi, R.E. (1991) Population dynamics of the cinnabar moth (*Tyria jacobaeae*): oscillations due to food limitation and local extinction risks. *Netherlands Journal of Zoology*, **41**, 158–73.

Morris, R.F. (1959) Single-factor analysis in population dynamics. *Ecology*, **40**, 580–8.

Morris, W.F. (1993) Predicting the consequences of plant spacing and biased movement for pollen dispersal by honey bees. *Ecology*, **74**, 493–500.

Murdoch, W.W. & Reeve, J.D. (1987) Aggregation of parasitoids and the detection of density dependence in field populations. *Oikos*, **50**, 137–41.

Nicholson, A.J. (1954) An outline of the dynamics of animal populations. *Australian Journal of Zoology*, **2**, 9–65.

Nicholson, A.J. & Bailey, V.A. (1935) The balance of animal populations. *Proceedings of the Zoological Society, Part 1. London*, **3**, 551–98.

Olmstead, I. & Alvarez Buyalla, E.R. (1995) Sustainable harvesting of tropical trees. *Ecological Applications*, **5**, 484–500.

Park, T., Leslie, P.H. & Mertz, D.B. (1964) Genetic strains and competition in populations of *Tribolium*. *Physiological Zoology*, **37**, 97–161.

Pearl, R. & Reed, L.J. (1920) On the rate of growth of the population of the United States since 1790 and its mathematical representation. *Proceedings of the National Academy of Science*, **6**, 275–88.

Perry, J.N. & Gonzalez-Andujar, J.L. (1993) Dispersal in a metapopulation neighbourhood model of an annual plant with a seed bank. *Journal of Ecology*, **81**, 453–63.

Pielou, E.C. (1977) *Mathematical Ecology*. John Wiley & Sons. New York

Pimm, S.L. (1982) *Food Webs*. Chapman & Hall, London.

Pimm, S.L. (1984) The complexity and stability of ecosystems. *Nature*, **307**, 321–6.

Pimm, S.L. & Lawton, J.H. (1977) The number of trophic levels in ecological communities. *Nature*, **268**, 329–31.

Pimm, S.L., Lee Jones, H. & Diamond, J. (1988) On the risk of extinction. *American Naturalist*, **132**, 757–85.

Pontin, A.J. (1982) *Competition and Coexistence of Species*. Pitman, London.

Prout, T. & McChesney, F. (1985) Competition among immatures affects their adult fecundity: population dynamics. *American Naturalist*, **126**, 521–58.

Reeve, J.D. & Murdoch, W.W. (1985) Aggregation by parasitoids in the successful control of the California red scale: a test of theory. *Journal of Animal Ecology*, **54**, 797–816.

Renshaw, E. (1991) *Modelling Biological Populations in Space and Time*. Cambridge University Press, Cambridge.

Ricker, W.E. (1954) Stock and recruitment. *Journal of the Fisheries Research Board of Canada*, **11**, 559–623.

Roberts, A. (1974) The stability of a feasible random ecosystem. *Nature*, **251**, 607–8.

Rogers, D.J. (1970) *Aspects of host–parasite interactions in laboratory populations of insects.* Unpublished D. Phil thesis. University of Oxford.

Root, R.B. (1967) The niche exploitation of the blue-grey gnatcatcher. *Ecological monographs*, **37**, 317–50.

Rosenzweig, M.L. & MacArthur, R.H. (1963) Graphical representation and stability condition of predator–prey interactions. *American Naturalist*, **97**, 209–23.

Schmid, P.E. (1992) Community structure of larval *Chironomidae* (*Diptera*) in a back water area of the River Danube. *Freshwater Ecology*, **27**, 151–67.

Segel, L.A. (1984) *Modeling Dynamic Phenomena in Molecular and Cellular Biology.* Cambridge University Press, Cambridge.

Seifert, R.P. & Seifert, F.H. (1976) A community matrix analysis of *Heliconia* insect communities. *American Naturalist*, **110**, 461–83.

Silander, J.A. & Antonovics, J. (1982) Analysis of interspecific interactions in coastal plant community – a perturbation approach. *Nature*, **298**, 557–60.

Silvertown, J. & Smith, B. (1989) Germination and population structure of spear thistle *Cirsium vulgare* in relation to experimentally controlled sheep grazing. *Oecologia*, **81**, 369–73.

Silvertown, J., Holtier, S., Johnson, J. & Dale, P. (1992) Cellular automation models of interspecific competition for space – the effect of pattern on process. *Journal of Ecology*, **80**, 527–34.

Solbrig, O.T., Sarandon, R. & Bossert, W. (1988) A density-dependent growth model of a perennial herb, *Viola fimbriatula*. *American Naturalist*, **131**, 385–400.

Soule, M. (1987) *Viable Populations for Conservation.* Cambridge University Press, Cambridge.

Species Survival Commission (1994). *IUCN Red List Caregories*, International Union for Conservationist Nature and Natural Resources Gland.

Steele, J. (1974) Stability of plankton ecosystems. In: *Ecological Stability* (eds M.B. Usher & M.H. Williamson) 179–91 Chapman & Hall, London.

Straw, N.A. (1991) *Report on Forest Research*. p. 43. HMSO, London.

Taylor, R.A.J. & Taylor, L.R. (1979) A behavioural model for the evolution of spatial dynamics. In: *Population Dynamics* (eds R.M. Anderson, B.D. Turner & L.R. Taylor), pp. 1–28. Blackwell Scientific Publications, Oxford.

Taylor, B.L. (1995) The reliability of using population viability analysis for risk classification of species. *Conservation Biology*, **9**, 551–8.

Tregonning, K. & Roberts, A. (1979) Complex systems which evolve towards homeostasis. *Nature*, **281**, 563–4.

Turchin, P. (1990) Rarity of density dependence or population regulation with lags? *Nature*, **344**, 660–3.

Turchin, P. & Taylor, A.D. (1992) Complex dynamics in ecological time series. *Ecology*, **73**, 289–305.

Usher, M.B. (1973) *Biological Management and Conservation*. Chapman & Hall, London.

Varley, G.C. (1947) The natural control of population balance in the knapweed gallfly. *Journal of Animal Ecology*, **16**, 139–87.

Varley, G.C. & Gradwell, G.R. (1960). Key factors in population studies. *Journal of Animal Ecology*, **29**, 399–401.

Varley, G.C. & Gradwell, G.R. (1963) Predatory insects as density-dependent mortality factors. *Proceedings 16th International Congress of Zoology*, **1**, 240.

Verhulst, P.F. (1838) Notice sur la loi que la population suit dans son accroissement. *Correspondances Mathématiques et Physiques*, **10**, 113–21.

Volterra, V. (1926) Fluctuations in the abundance of a species considered mathematically. *Nature*, **118**, 558–60.

Volterra, V. (1927) see Lotka (1927).

Volterra, V. (1928 and 1931) Variations and fluctuations of the number of individuals in animal species living together. *Journal du Conseil international pour l'exploration de la mer*, III, vol. 1, pp. 3–51.

(Translation of 1926 Variazione e fluttuazioni del numero d'individui in specie animali conviventi. *Memoria della R. Accademia Nazionale dei Lincei (ser 6.)*, **2**, 31–113.) Reprinted in Chapman, R.N. (1931) *Animal Ecology*, pp. 409–48. McGraw-Hill. New York and London.

Watkinson, A.R. (1980) Density-dependence in single-species populations of plants. *Journal of Theoretical Biology*, **83**, 345–57.

Watkinson, A.R. (1987) Plant population dynamics. In: *Plant Ecology* (ed. M.J. Crawley), pp. 137–84. Blackwell Scientific Publications, Oxford.

Wilson, E.O. & Bossert, W.H. (1971) *A Primer of Population Biology*. Sinauer Associates, Sunderland, Mass.

Wilson, J.B. & Roxburgh, S.H. (1992) Application of community matrix theory to plant competition data. *Oikos*, **65**, 343–8.

Woiwod, I.P. & Hanski, I. (1992) Patterns of density-dependence in moths and aphids. *Journal of Animal Ecology*, **61**, 619–30.

Wolfram, S. (1984) Cellular automata as models of complexity. *Nature*, **311**, 419–24.

Wyatt, T. (1974) Red tides and algal strategies. In: *Ecological Stability* (eds M.B. Usher & M.H. Williamson). 35–40 Chapman & Hall, London.

Zhang, J. & Zheng, G. (1990) Numbers and populations structure of Cabot's Tragopan. In: *Pheasants in Asia* (eds Hill, D.A., Garson, P.J. & Jenkins, D.). World Pheasant Association, Reading, UK.

Index